设计无界
世界设计之都大会主题展

Design Beyond
The Theme Exhibition of the
World Design Cities Conference

2022—2023

同济大学出版社·上海
TONGJI UNIVERSITY PRESS · SHANGHAI

娄永琪 Yongqi Lou
编著

前言
Preface

上海这座城市是最好的设计大学

上海自2010年加入联合国教科文组织（UNESCO）创意城市网络"设计之都"以来，不断汇聚创意与设计人才和资源，推动了产业和社会的蓬勃发展，也让上海的全球科创中心和卓越城市建设气象日新。2022年1月，上海制定发布了《上海建设世界一流"设计之都"的若干意见》；9月15日，首届世界设计之都大会开幕，成为继世界人工智能大会后，对上海科创中心建设又一具有战略意义的国际会议。这也是这座伟大城市向全世界发出的通过设计驱动创新发展、推动经济转型升级和人民城市建设的宣言书。同济大学作为大会的主要承办方，全方位参与了大会的策划和举办，我本人也连续两届被任命为大会的创意总监，负责大会学术和创意的整体协调工作。

我认为作为官方主推的年度设计盛事，世界设计之都大会以其权威高度、学术深度、文化态度、产业广度、国际化程度有别于国内的一众设计类活动。其中，思想的引领既是大会的重要目标，同时也是大会成功的重要保障。作为有全球影响力的世界设计之都，上海应该整合全球视野、前瞻思想和区域社会经济发展，通过出观点、出思想、出作品、出品牌、出产业，逐步凝结一个不断壮大、主动担当、引领未来的创意社群，为上海、中国和世界的可持续发展作出贡献。

首两届大会，分别以"设计无界，相融共生"和"设计无界，造化万象"为主题。为了更好地从学术角度演绎这次大会，我牵头策展了两届相互关联的主题展，分别是"人·人，人·自然"和"创意社群：嵌入式、可感知与高交互"。第一年的主题展，事实上是一次面向全社会的设计启蒙。我们将设计置于一个宏大的视角之下，讨论当今设计的新角色、新使命、新对象和新方法。人类创造了"人造物世界"，以更好地调适人与自然之间的关系，实现高质量的生存和生活，这个人造物世界，现在已经成为人类文明的载体。在此过程中，设计始终是人类有意识创造的先导。人们创造环境，而环境反过来也影响人们生活的方方面面。比如，作为人造物世界的新物种——数字网络世界，尽管诞生至今不过80年，但已经全面嵌入我们生活的每个片段。在"技术"的全面影响下，人类很容易陷入一种技术崇拜，而忘记了人类存在的根本意义。这时候，只有重新回到生活世界，去问一些与人类福祉密切相关的大问题，并运用设计思维和设计实践来提供创新解决策略，才是人类价值的实现路径。

2022年世界设计之都大会的主题展，是希望在全世界大流行的新冠疫情的尾声，探讨一系列"人与人"以及"人与自然"关系的大问题，包括食物、健康、信任、环境、技术和繁荣，以及展现设计如何应对这些问题的全球行动。从展览中，我们可以看到，当今的设计早已不是一种美学工具，而是化身为新质生产力的核心创新要素。设计正在通过对"意义和方式"的重新定义，全面赋能科技革命性突破、生产要素创新性配置、产业深度转型升级，推动生产关系的质变和全要素生产率提升。这种创新，按理查德·布坎南所说的设计四秩序来解析，既需要有宏观尺度的系统、环境和机制的设计，也需要有服务、流程和交互的非物质设计，更需要落到产品和符号传达层面，方可被人感知。我们可以看到，大部分成功的设计都是针对问题，综合运用以上各种秩序的设计的结果。哪里有问题，哪里就需要设计。问题越严重，设计的价值和贡献也就越大。不论是发现问题、定义问题和解决问题，设计和设计思维都是能否产生突破性创新的关键。

2023年的主题展，我们将注意力集中到人的身上，特别是一个城市的创意社群。人是"人—自然—人造物—赛博"四系统中，唯一具有真正的主观能动性的系统力量。如何凝结一个积极的、开放的、协作的创意社群，是任何设计驱动创新和美好城市建设实现的关键。上海举办世界设计之都大会，就是要催生和支持这个社群的成长，并为身处之中的人们提供一个思想交流、创意互鉴、洞察未来的全球共创共享生态。以人工智能、移动互联网、云科技、生命科学、量子计算为代表的现代科技发展，正在全方位地重新定义人们的工作、生活、创新和学习方式，三者之间的界限也逐渐模糊。对城市而言，各种城市功能分区及以其为基础配置资源的传统方式遭遇了前所未有的挑战。学校校区、居住社区、产业园区、商业街区等传统城市功能区的界限也正在变得模糊。工作不一定发生在单位里，学习不一定发生在学校里，创新也不一定在实验室里……全新的城市空间组织形态、社会经济运行模式和创新转化范式呼之欲出。

这些新可能，也为解决很多城市发展的积弊带来了机会。比如随着世界城市化率越来越高，大部分城市也越来越同质化。我认为一个城市的魅力，往往不体现在大马路和地标区域上。一个有持久吸引力的城市，一定是一个内涵丰富的城市。小马路上、巷子里、社区里有没有活力、动力、魅力才是关键。因此，2023年的主题展，我以"创意社群：嵌入式、可感知与高交互"为题，把自2015年开始，我推动同济大学设计创意学院和杨浦区四平路街道合作，在周边社区逐步建起来的一系列"小而互联"的实验室群落，用高度抽象的方式搬进了大会的展览。每一个盒子都代表一个实验室，盒子里和盒子之间，都是各种生动的人与人之间交互和协同创新的故事。通过这种社区的方式，呈现上海这座城市里随处可见的设计作品、设计事件和设计力量，传达上海"城市处处有设计""寓设计于生活"以及"人人都是设计创新者"的理念、主张和行动。城市社区和市井的烟火气，让城市创新有了人的温度，也给创新贴上了时代、地域、文化的个性化标签。在展览期间，参展实验室轮流每半天举办一次工作坊，高度还原了这个城校共建的融于城市街区的创意社群的活力。

我一直认为"城市是最好的大学，上海这座城市就是最好的设计大学"。上海世界设计之都建设的底层逻辑，不是基于全球城市之间的竞争。作为世界设计之都，上海这所"城市设计大学"，不仅要充分发挥一流人才密集、产业要素集聚、应用场景开放、多元文化交融、国际合作广泛等优势，将设计全面融入城市的产业发展、空间环境、公共服务、民众生活和城市品牌建设的方方面面，而且应该为全世界通过设计创意推动创新转化和可持续发展提供知识、人才和文化的支持。特别是如何通过设计驱动，充分发挥上海链接中国和世界、当下和未来的"结构洞"效应和"创新策源实验室"效应，这将是上海世界设计之都和全球科创中心建设的重要任务。

本书的编辑和出版，既是对过去两年大会的总结，也是对新一届大会的献礼。本书收录了首两届世界设计之都大会的主题展图文册，以及大会的主视觉设计、总体规划、展览和会场设计等内容。我希望通过这本书，记录下过去两年筹备世界设计之都大会的工作，在此过程中对设计、创新、设计产业、设计教育的发展，以及对上海世界设计之都和科创中心建设的一些思考。我要感谢上海市各级领导的支持，要感谢张磊、章明、王敏、杜钦、张屹南、周洪涛、刘刊、柳喆俊、刘畅、苏运升、周晴、安东·西比克等来自同济大学的设计、规划、展览、论坛、会务和国际合作团队的全身心投入，感谢连续两届大会主题展的国际策展团队，感谢为展览提供展品的合作单位和个人，也要感谢同济大学出版社和本书编辑团队的大力支持。有了大家的协同共创，大会的顺利举办和本书的出版才成为可能！

在本书的编辑过程中，2024 年世界设计之都大会正在紧锣密鼓地筹备，今年的主题是"设计无界，新质生长"。大会将对如何创造性地培育和运用"设计"这一重要的新质生产力创新要素展开充分讨论，并积极探索面向未来经济、社会和环境可持续发展的全新范式。我们相信，上海这座城市将继续以它无尽的创造力和不懈的追求，塑造一个既属于当下、又通向永恒的未来。那些今日的梦想、创意与设计，终将在明日的光辉中化为新的传奇！

娄永琪 教授
同济大学副校长
英国皇家艺术学院荣誉博士
瑞典皇家工程科学院院士
2024 年 8 月 15 日

Shanghai Is the Best Design University

In 2010, Shanghai joined the UNESCO Creative Cities Network as City of Design, demonstrating its determination to put design at the core of its sustainable economic growth and development plan. Since then, the city has continuously attracted creative talents and design resources, combining modern design and cultural heritage, and driving growth in both industry and society in the context of the global innovation center initiatives. In January 2022, Shanghai issued the "Opinions on Building a World-Class 'City of Design'", setting forth ambitious goals for its development in the design sector. This was followed by the inaugural World Design Cities Conference on September 15, 2022. This conference, held after the World Artificial Intelligence Conference, holds strategic significance for Shanghai's development as a science and innovation hub. It represents the city's commitment to leveraging design as a catalyst for innovation, economic transformation, and urban development. Tongji University, as the primary organizer of the conference, played a crucial role in its planning and execution. I was privileged to be appointed as the creative director for two consecutive terms, overseeing the academic and creative coordination of the event.

As the officially endorsed annual design event, the World Design Cities Conference stands out from other domestic design activities due to its authoritative stature, academic level, cultural diversity, industrial representativeness, and international impact. Thoughts Significance is both a key objective of the conference and a crucial factor for its successful execution. As a globally influential "City of Design", Shanghai should integrate a global perspective, forward thinking, and regional socio-economic development. By generating insights, ideas, works, brands, and industries, Shanghai is committed to cultivate a robust, proactive, and forward-looking creative community which will contribute to the sustainable development of the city, the country, and the rest of the world.

In the first two conferences, themed "Vision in Perspective: Design Beyond Borders, Diversity and Togetherness" and "Design Beyond Creativity", I led the curation of two interconnected theme exhibitions titled "People to People, Human to Nature" and "NICE Commune: Embedded, Sensible and Interactive". These exhibitions aimed to explore and present the conference's themes from an academic perspective. The first thematic exhibition served as a public-oriented design enlightenment initiative to indicate the latest development of design. We placed design within a grand perspective, examining its evolving roles, missions, objects, and methods in contemporary contexts. Human being has created an "artificial world" to optimize the interaction with nature and achieve high-quality existence and living. This artificial world has now become a vessel of human civilization. Throughout this process, design has consistently been the vanguard of human intentional creativity. People create environments, and these environments, in turn, influence every aspect of our lives. For instance, the cyber system—a new species of the artificial world—has been fully integrated into every fragment of our daily lives despite its mere 80-year existence. Amid the pervasive influence of technology, humanity tends to fall into a kind of technological worship, potentially overlooking the fundamental meaning of human existence. In such times, returning to the life-world and addressing critical questions closely related to human well-being, while applying design thinking and solutions to provide innovative strategies, becomes the pathway to realizing human values.

The 2022 World Design Cities Conference Theme Exhibition aimed to explore a series of critical issues concerning "People to People" and "Human to Nature" relations in a global scale, particularly in the aftermath of the global COVID-19 pandemic. These global challenges included food, health, trust, environment, technology, and prosperity, while showcasing how design addresses these global challenges. The exhibition revealed that contemporary design has transcended its role as merely an aesthetic tool, evolving into a core element of innovative productivity. Through redefining "meaning and means", design is now empowering technological breakthroughs, innovative configurations of production factors, and revolutionary industrial transformations. It is driving qualitative changes in production relations and enhancing total factor productivity. According to Richard Buchanan's framework of the four orders of design, the involvement of design in this process requires not only macro-level system, environment, and mechanism design but also immaterial design involving services, processes, and interactions. Additionally, it necessitates product and communication design that can be perceived by people. Most successful designs are results of addressing problems through the integration of these various design orders. The presence of a problem indicates the need for design, and the more severe the problem is, the greater the value and contribution of design is. Whether it is identifying, defining, or solving problems, design and design thinking are crucial for achieving breakthrough innovations.

For the 2023 Theme Exhibition, the focus shifted to people, particularly the creative communities. Humans are the only system with genuine subjective agency among the four systems of "Human-Nature-Artifacts-Cyber". Building a positive, open, and collaborative creative community is crucial for achieving design-driven innovation and enhancing urban development for any creative city. Shanghai's hosting of the World Design Cities Conference aims to foster and support the growth of this community, providing a global co-creation and sharing ecosystem for intellectual exchange, creative collaboration, and future insights. Modern technological advancements, including artificial intelligence, mobile internet, cloud computing, life sciences, and quantum computing, are redefining how people work, live, innovate, and study, with boundaries between these areas becoming increasingly blurred. For cities, traditional methods of functional zoning and resource allocation face unprecedented challenges. The distinctions between school campuses, residential communities, industrial parks, and commercial districts are becoming less clear. Work is no longer confined to offices, learning is not restricted to schools, and innovation does not exclusively occur in laboratories. New forms of urban spatial organization, socio-economic models, and innovation transformation paradigms are emerging.

These new possibilities also present opportunities to address many of the long-standing issues in urban development. For instance, as the global urbanization rate increases, many cities are becoming more homogenized. I believe that a city's charm is often not reflected in its main streets and landmark areas. A city with lasting appeal is one with rich inner content. The vitality, dynamism, and charm of small streets, alleys, and communities are crucial. Therefore, for the 2023 thematic exhibition, I chose the theme "NICE Commune: Embedded, Sensible, and Interactive" to showcase a series of "small but interconnected" laboratory clusters that I have been developing since 2013 at College of Design and Innovation of Tongji University in collaboration with the Siping neighborhood in Yangpu District. These clusters were abstractly represented in a format of "box community" in the exhibition. Each box symbolizes a laboratory, and within and between these boxes are various vibrant interaction and collaboration among different stakeholders. This community-based presentation highlights the design works, events, and forces prevalent throughout Shanghai, conveying the city's concept that "Design is everywhere", "Design is embedded in life", and "Everyone is a design innovator". The lively atmosphere of urban communities and city scenes adds a human touch to urban innovation, personalizing it with the tags of the era, region, and culture. During the exhibition, each participating laboratory hosted a workshop every half day, vividly replicating the vitality of this city-university co-created creative community embedded in urban neighborhoods.

I have always believed that "The city is the best university, and my dream is to shape the city of Shanghai to be the best design university." The underlying logic of Shanghai's development as a World Design City is not based on competition among global cities. Shanghai, as a "city design university", must leverage its advantages, including a concentration of top talent, accumulation of industrial elements, open application scenarios, diverse cultural integration, and extensive international cooperation. Design should be fully integrated into all aspects of urban development, from industry and spatial environment to public services, daily life, and city branding. Moreover, Shanghai should offer global support in knowledge, talent, and culture, driving innovation and sustainable development through design creativity. In particular, it will be a crucial task to effectively maximize Shanghai's roles both as a "structural hole" which linking China with the rest of the world and the "innovation cradle" in a global scale.

The editing and publication of this book serve both as a summary of the two past conferences and as a tribute to the upcoming edition. This book includes the thematic exhibition catalogs from the first two World Design Cities Conferences, along with content related to the conference's main visual design, site planning, exhibition, and venue design. I hope this book captures the efforts involved in preparing for the World Design Cities Conference over the past two years, along with reflections on the development of design, innovation, the design industry, design education, as well as the evolution of Shanghai as a Design City and innovation hub. I would like to express my gratitude to the leaders of Shanghai at various levels for their support, to Lei Zhang, Ming Zhang, Min Wang, Qin Du, Yinan Zhang, Hongtao Zhou, Kan Liu, Zhejun Liu, Chang Liu, Yunsheng Su, Qing Zhou, Aldo Cibic, and other colleagues from Tongji University involved in design, planning, exhibitions, forums, administration affairs, and international cooperation for their wholehearted dedication, to the international curatorial teams for the thematic exhibitions of the past two years, and to the collaborating units and individuals who provided the works for the exhibitions. I also extend my thanks to Tongji University Press and the editorial team of the book for their strong support. It is through everyone's collaborative effort that the successful hosting of the conference and the publication of this book have become possible!

As this book was being edited, preparations for the 2024 World Design Cities Conference were proceeding at full speed. This year's theme is "Design, Reframing Growth". The conference will thoroughly discuss how to creatively cultivate and utilize design as a key element of productivity innovation and will actively explore new paradigms for future economic, social, and environmental sustainability. We believe that Shanghai will continue to shape a future that is both contemporary and timeless through its boundless creativity and relentless pursuit. The dreams, ideas, and designs of today will, in the brilliance of tomorrow, transform into new legends!

Prof. Yongqi Lou

Vice President of Tongji University

Honorary Doctor of Royal College of Art

Fellow of Royal Swedish Academy of Engineering Sciences

August 15, 2024

目录
Contents

前言 | 上海这座城市是最好的设计大学　　　　　i

专访 | 上海举办世界设计之都大会背后有何深意？　　1
娄永琪这样说……

第一部　2022年首届"设计无界，相融共生"世界设计之都大会主题展——人·人，人·自然

策展团队	16
策展序言	20
策展结构：六大叙事	22
作品介绍	24
主题1：食物	25
主题2：健康	34
主题3：信任	44
主题4：环境	60
主题5：技术	81
主题6：繁荣	90
展览设计	123
展览实景	131
工作坊	145
致谢	152

第二部　2023年"设计无界，造化万象"世界设计之都大会主题展——创意社群：嵌入式、可感知与高交互

策展团队	158
策展序言	162
策展结构：嵌入社区的创新实验室集群	164
实验室介绍	166
实验室参展作品	178
设计生态集锦展区	224
展览设计	242
展览实景	246
工作坊	259
致谢	268
WDCC设计花絮	269
世界设计之都大会主视觉设计	270
世界设计之都大会开幕秀：设计交响（2022）	274
世界设计之都大会开幕秀：化（2023）	278
世界设计之都大会"人民城市 | 处处有设计"建筑展（2023）	282
世界设计之都大会会场设计（2022—2023）	292
世界设计之都大会主题馆：无止园（2022）	300
世界设计之都大会论坛场景设计（2023）	306
世界设计之都大会演讲台设计	314

Preface | Shanghai Is the Best Design University   i

Exclusive Interview | The Meaning Behind Shanghai Hosting the World Design Cities Conference: Insights from Yongqi Lou   1

Part One

The 2022 Inaugural "Vision in Perspective: Design Beyond Borders, Diversity and Togetherness" World Design Cities Conference Theme Exhibition—People to People, Human to Nature

Curatorial Team	16
Curatorial Statement	20
Curatorial Structure: Six Major Narratives	22
Collections	24
Theme 1: Food	25
Theme 2: Health	34
Theme 3: Trust	44
Theme 4: Environment	60
Theme 5: Technology	81
Theme 6: Prosperity	90
Exhibition Design	123
Exhibition Scenes	131
Workshops	145
Acknowledgments	152

Part Two

The 2023 "Design Beyond Creativity" World Design Cities Conference Theme Exhibition— NICE Commune: Embedded, Sensible and Interactive

Curatorial Team	158
Curatorial Statement	162
Curatorial Structure: Innovation Labs Embedded in Communities	164
Labs' Introduction	166
Introduction of Labs' Works	178
Design Ecological Exhibition Area	224
Exhibition Design	242
Exhibition Scenes	246
Workshops	259
Acknowledgements	268
Design Sidelights of WDCC	269
Main Visual Identity System Design of WDCC	270
Opening Show of WDCC 2022: Design Symphony	274
Opening Show of WDCC 2023: Metamorphosis	278
"People's City	Design Everywhere" Architectural Exhibition of WDCC 2023	282
Site Planning of WDCC 2022-2023	292
WDCC 2022 Theme Pavilion: Endless Garden	300
Scenography Design of the Conference Venue of WDCC 2023	306
Design of the WDCC Main Podium	314

专访
Exclusive Interview

上海举办世界设计之都大会背后有何深意？娄永琪这样说……*

The Meaning Behind Shanghai Hosting the World Design Cities Conference: Insights from Yongqi Lou

乔梦婷

Mengting Qiao

* 本文原文发表于《上观新闻》（2022 年 9 月 13 日）

This article was originally published in *Shangguan News* (September 13, 2022), by Mengting Qiao.

自 2010 年上海加入联合国教科文组织"创意城市网络"，定名为"设计之都"至今，上海"设计之都"建设已然迈过十余个年头。在 2022 年 2 月 17 日举行的推进大会上，时任市委副书记、市长龚正按下启动按钮，标志着世界一流"设计之都"建设开启了新的篇章。

面对即将召开的首届世界设计之都大会，娄永琪作为大会幕后的策划者之一，将如何向公众描绘这场大会及其所释放的信号与未来方向？

采访从大会名称中的"世界设计之都"开始，聊到了他对这次大会的期待、对大会内容的剧透、对设计人才培养的看法、对上海设计产业发展的见解等。对话中，娄永琪不时提到同样在上海举办的"世界人工智能大会"，提到上海建设全球影响力的"科创中心"的宏伟目标。在他看来，一手抓科技，一手抓创意，正是未来上海卓越城市建设的核心驱动力。本次大会是上海的重要选择，更是上海如何不断超越自己的一个答案。

Since Shanghai joined the UNESCO Creative Cities Network in 2010 and bore the title of "City of Design", the development of Shanghai as a "City of Design" has spanned over one decade. On February 17 2022, during the promotion conference, the Mayor Zheng Gong pressed the start button, marking the beginning of a new chapter in the building of a world-class "City of Design".

At that time, in anticipation of the upcoming inaugural World Design Cities Conference, how will Prof. Yongqi Lou, one of the key planners behind the conference, articulate to the public the signals and directions of the conference?

Our interview began with a discussion on the term "World Design Cities" in the conference title and then moved on to his expectations for the conference, insights into the content, views on design talent development, and perspectives on the development of Shanghai's design industry. Throughout the conversation, Prof. Lou frequently referenced the "World Artificial Intelligence Conference" also held in Shanghai and the city's ambitious goal of becoming a global "Innovation Center". In his view, simultaneously focusing on technology and creativity is the core driving force for Shanghai pursuing her future excellence. This conference is a significant choice for Shanghai and also an answers to how it continually surpasses itself.

2022 年世界设计之都大会官方网站
2022 World Design Cities Conference Official Website

01 背后是上海更大的雄心

问：您心目中的"世界设计之都"是怎样的？上海对此有着怎样的表现与诠释？

娄永琪：任何一座城市，但凡有雄心建设"世界设计之都"，就意味着设计将成为这座城市个性、文化、品牌的重要组成。当人们想起这座城市的若干特征时，至少有一个不可忽略的特征是"设计"。当然，作为"世界设计之都"，它也必须构筑对全球设计的影响力，影响设计领域的世界格局、影响设计产业的未来发展。

2010 年，上海加入联合国教科文组织创意城市网络，被命名为"设计之都"，这是上海世博会举办期间上海的重要战略选择。而建设世界一流"设计之都"，是 2022 年上海市委市政府对于新阶段新目标的重要部署，与初始阶段相比，这背后是上海更大的雄心。

值得一提的是，今年与大会同月举办的世界人工智能大会，已连续举办了五届，令人印象深刻。无论是世界人工智能大会，还是世界设计之都大会，背后都是上海高质量发展的重要举措，可以说是上海卓越城市建设的"一体两翼"。

世界设计之都大会对上海而言，绝不仅仅是多了一个大会，它是上海建设世界一流"设计之都"的重要载体。这也是上海对于"如何解决下一阶段高质量发展"的重要选择，也是上海如何不断超越自己的一个答案。

01 Shanghai's Ambition

Q: What is your vision of a "World Design City", and how does Shanghai embody and interpret this concept?

Lou: Any city that aspires to be a "World Design City" means that design will become a crucial part of its identity, culture, and brand. When people think of this city, at least one prominent feature should be "design". Furthermore, as a "World Design City", it must also establish global influence, affecting the world design landscape and shaping the future development of the design industry.

In 2010, Shanghai joined the UNESCO Creative Cities Network and was designated as a "City of Design", marking a significant strategic decision for the city during the Expo. The goal of developing a world-class "City of Design" represents a major initiative by the Shanghai Government in 2022 to meet new stages and objectives, reflecting a broader ambition.

It is worth noting that the World Artificial Intelligence Conference, held in the same month as this year's conference, has been held impressively for five consecutive years. Both the World Artificial Intelligence Conference and the World Design Cities Conference are integral to Shanghai's high-quality development strategy and can be considered as the "two wings" of Shanghai's pursuit of excellence.

For Shanghai, the "World Design Cities Conference" represents more than just an additional event; it is a crucial vehicle for the city's development as a world-class "City of Design". It is an important choice for Shanghai in addressing the challenge of achieving high-quality development at the next stage and serves

至于设计之都建设成功的标志,我想提一个比较感性的指标:如果未来有一天,人们想了解全世界最新的设计思想和潮流时,首先想到的是去上海看看,就像如今人们看设计会想到去纽约、米兰、东京、巴黎、伦敦一样;未来有一天想了解中国和世界最前沿的设计,如果只能去一个地方看,去哪里?这个答案是否就是上海?如果这个场景可以实现,可以说建设世界一流"设计之都"的宏伟目标就实现了。我真心地认为中国这么多城市中,只有上海最有可能最先达到这么一个目标。

问:上海为何要在当下这个时期举办这样一场大会?它将释放怎样的信号?

娄永琪:今年上海正式提出要加快建设具有世界影响力的社会主义现代化国际大都市,努力使"世界影响力"的能级显著提升、"社会主义现代化"的特征充分彰显、"国际大都市"的风范更具魅力。如何推进落实这些高瞻远瞩的发展战略?我想,"创意设计"和"人工智能"一样,会是上海这座城市的独特优势。

受全球疫情影响,这次举办首届世界设计之都大会也难免受到诸多客观条件限制,如不少国际与会嘉宾无法亲临现场等。但这样一场由政府主导发起、迄今少有的高规格的设计大会,必将成为一次令人振奋的集结号——吹响上海建设世界一流"设计之都"的号角,传达上海对人类文明智慧、想象能力、创造能力的尊重与推崇,设计将成为未来上海科创中心建设的重要内涵和抓手。与此同时,世界设计之都大会也必将成为上海创新型城市建设的重要品牌与名片。

问:对标2030年上海全面建成世界一流"设计之都"的目标,此次大会扮演怎样的角色?对上海城市建设与上海设计产业发展将带来何种影响?

娄永琪:如果说城市是文化的容器,那么世界设计之都大会将以设计为媒,推升上海这个超级"文化的容器"变得更加包容,更加有吸引力。在此过程中,城市对人才的吸引力得到了提升——不仅是设计人才,一座崇尚创意的城市,也一样会吸引更优秀的科学家、工程师、人文学者、艺术家、工程师、投资人等多领域的创新人才,让城市的创新生态极大提升;同时,作为人民城市建设的重要一环,大会也将极大提升城市对设计的理解,提升全城设计氛围,让市民不仅每时每刻都感受到"设计点亮生

as an answer to how the city can continually surpass itself.

As for the indicators of successful development as a City of Design, I would like to propose a relatively subjective measure: if, in the future, people think of Shanghai as the first place to explore the latest global design ideas and trends — much as people currently think of New York, Milan, Tokyo, Paris, and London for design — then this would signify success. Similarly, if Shanghai becomes the go-to place for understanding the most cutting-edge design in China and the rest of world, the ambitious goal of becoming a world-class "City of Design" would be realized. I sincerely believe that, among many cities in China, Shanghai is the most likely to achieve this standard first.

Q: Why is Shanghai holding such a conference at this particular time, and what signals will it send?

Lou: This year, Shanghai has officially announced its goal to become a socialist modern international metropolis with global influence. The aim is to significantly enhance the level of "global influence", fully highlight the characteristics of "Chinese modernization", and make the "international metropolis" more attractive. How to implement these forward-looking development strategies? I believe that "creative design", like "artificial intelligence", will be a unique advantage for the city of Shanghai.

Given the global pandemic, the inaugural World Design Cities Conference will inevitably face various objective constraints, such as the inability of many international guests to attend in person. However, this high-profile design conference, initiated and led by the government, will undoubtedly serve as an inspiring rallying call — marking the beginning of Shanghai's journey to becoming a world-class "City of Design", and reflecting Shanghai's respect and admiration for human civilization's wisdom, imagination, and creativity. Design will become an essential component and focus in the future development of Shanghai's innovation center. At the same time, the World Design Cities Conference will also become an important brand and calling card for Shanghai's development as an innovative city.

Q: In relation to Shanghai's goal of becoming a world-class "City of Design" by 2030, what role does this conference play? What impact will it have on Shanghai's urban development and design industry?

Lou: If a city is considered as a container of culture, the World Design Cities Conference will act as a medium that elevates Shanghai, this super "cultural container," making it more inclusive and attractive. This process includes enhancing the city's appeal to talent — not only design professionals but also scientists, engineers, humanists, artists, and investors, thus significantly boosting the city's innovation ecosystem. Moreover, as

活"的精彩，更深深参与其中，让设计成为"无中生有、点石成金、莫名其妙"的上海城市创新推进器。

讲到设计与产业的关系，我觉得其关系就像是酵母与面粉一般。一方面，这次大会必将推动上海创意设计资源能级得到极大提升，强化"酵母"的功能。但对上海这座城市而言，仅集聚一流的设计创意资源是不够的，更重要的是酵母如何与面粉一起，在合适的容器中，配以合适的温度、水，发出更优质的面团，做出更美味的面包。因此，设计产业一方面需要一流的创意创新，并以此为催化剂；另一方面也需要全方位地与城市、产业充分融合，只有这样，创新转化才能够得以实现。

问：放眼全球，此次大会将对全球设计生态体系发挥怎样的引领作用？

娄永琪：2022 年世界设计之都大会为期四天，由于疫情，活动不可避免地受到影响。在有限的时间和不确定的全球环境中，举办一场国际大会并非易事。

但上海决定举办这场大会，吹响这个集结号，就足以显示这座城市的底气、活力与担当，它对世界设计的影响将不可估量。这个影响不在这个会本身，而是作为一座影响世界创新、文化、经济走向举足轻重地位的城市——上海，在此刻选择了设计，这个信号本身的价值就远远超越了活动本身。因为大家会记住一个信号，而不是这颗具体的信号弹或是某支具体的号角。

02 演绎"设计无界，相融共生"

问：作为首届大会，2022 年世界设计之都大会的亮点与焦点在何处？您对此有何期许？

娄永琪：本次大会组委会格外重视大会筹办工作，尽管面临种种限制，我们仍将矢志不渝地秉持大会国际化、高规格水

a crucial component of building a "People's City", the conference will greatly enhance the city's understanding of design and elevate the overall design atmosphere. It will enable citizens to experience the brilliance of "design illuminating life" continuously and engage deeply, turning design into an essential driver of Shanghai's urban innovation.

Regarding the relationship between design and industry, I liken it to the relationship between yeast and flour. On one hand, the conference will significantly enhance Shanghai's creative design resources, amplifying the "yeast" function. However, for Shanghai, it is not enough to merely accumulate top-notch design resources. Equally important is how this "yeast" interacts with the "flour" in the right environment, with appropriate temperature and water, to produce high-quality dough and then delicious bread. Therefore, while the design industry needs outstanding creativity and innovation as catalysts, it also requires comprehensive integration with the city and industry. Only through such integration can innovation be effectively transformed.

Q: Looking globally, what role will this conference play in leading the global design ecosystem?

Lou: The 2022 World Design Cities Conference spans four days. Due to the pandemic, the event has inevitably been impacted. Organizing an international conference under such limited time and uncertain global conditions is a challenging task.

However, Shanghai's decision to host this conference and to issue this call to action demonstrates the city's confidence, vitality, and responsibility. Its impact on global design will be immeasurable. The influence lies not in the event itself but in the fact that Shanghai—a city crucial to the direction of global innovation, culture, and economy—has chosen to focus on design at this moment. The value of this signal far exceeds the event itself, as people will remember the message rather than the specific event or particular call to action.

02 Interpreting "Vision in Perspective: Design Beyond Borders, Diversity and Togetherness"

Q: As the inaugural conference, what are the highlights and focal points of the 2022 World Design Cities Conference? What are your expectations for it?

Lou: The organizing committee of this conference has placed

准、引导、探讨、碰撞、融汇一系列有思考、有主张、有分量的设计智慧，向世界传达上海的设计之道。为此，同济大学作为承办单位之一，与上海市经济和信息化委员会等单位紧密协同，策划了很多亮点，此处先为大家剧透其中三点。

其一，大会将推出一个"主题展"，和品牌展相映成趣。这个展览将如何体现关于设计发展的上海主张、上海高度？回溯人类历史上重要的展览，我们会发现好的展览绝不仅仅是"一场会""一个展"的概念，而是会留下深刻的思想与文化遗产，长久地影响人类文明的进步。为此，这次主题展就是希望能从"人·人，人·自然"的"关系"角度切入，聚焦"食物""健康""信任""环境""技术""繁荣"六大全球性议题，对大会主题"设计无界，相融共生"进行充分演绎，展现上海这座城市对于设计的思考与态度，同时也形成一个更持久地展开这些议题讨论的创意社群，从思想和行动上不断支持这个城市的创意涌现。

其二，大会将启动"国际设计百人"尖峰论坛，基于大会平台，邀请全球范围内一百个对设计有影响、有贡献、愿意支持上海设计之都建设的重量级设计人物，组成设计智库，为上海未来设计发展建言献策。当然，未来这一数量也将远远超过"一百"这个数字。我们希望这个平台能不断吸引愈来愈多的来自世界各地的设计人才加入上海的设计之都建设，并通过这个平台增强与上海这座城市的黏性。我相信，会有更多的创意火花将在此迸发，也会有更多的影响世界的设计思想、设计成果和设计人物在此诞生。

其三，大会将推出面向全球、分量十足的重要设计奖项——前沿设计创新奖。这个奖将分为三大类：一是杰出贡献奖，是奖励作出转折性贡献、开拓领域发展前沿的人，相当于设计界的终身成就奖；二是年度创新奖，奖励近年来对设计有独到理解，影响巨大，并且引领了未来设计方向的设计作品；三是未来创新奖，是全世界第一个面向K12青少年的设计奖项。

除此之外，上海市政府也在积极发动各个区县、高校、机构举办丰富多彩的全城响应活动，应该说，为期四天的大会只是一个开始，而上海全城的响应与持续的影响力，将是跨年的。

significant emphasis on its preparation. Despite various constraints, we remain committed to maintaining an international and high-standard level for the event, guiding, exploring, colliding, and integrating a series of thoughtful, substantive design insights, and conveying Shanghai's design philosophy to the world. To this end, Tongji University, as one of the hosts, has closely collaborated with other entities to plan several highlights. Here are three key aspects to preview.

Firstly, the conference will feature a "Theme Exhibition", which will complement the brand exhibitions. This exhibition aims to reflect Shanghai's perspectives and standards on design development. Reflecting on significant historical exhibitions, we find that impactful exhibitions are not merely "an event" or "an exhibition" but leave a profound intellectual and cultural legacy, significantly influencing the progress of human civilization. Therefore, this thematic exhibition seeks to approach the topic of "People to People, Human to Nature" from the relational perspective. It will focus on six global issues: "Food", "Health", "Trust", "Environment", "Technology", and "Prosperity", thoroughly interpreting the conference theme of "Vision in Perspective: Design Beyond Borders, Diversity and Togetherness". This exhibition will showcase Shanghai's contemplation and attitude toward design and also foster a creative community that continuously supports and develops these discussions, both intellectually and practically.

Secondly, the conference will launch the "International Design 100" summit forum, leveraging the conference platform to invite one hundred influential and contributing global design figures who are committed to supporting the development of Shanghai as a City of Design. These individuals will form a design think tank to provide strategic advice for the future design evolution of Shanghai. It is anticipated that this number will significantly increase in the future. We aim for this platform to continually attract an ever-growing number of design talents from around the world to participate in the development of Shanghai as a City of Design, thereby strengthening their connection with the city. I believe that this will spark more creative ideas and lead to the emergence of influential design concepts, achievements, and personalities.

Thirdly, the conference will introduce a significant global design award—the Frontier Design Innovation Award. This award will be categorized into three main categories: First, the Outstanding Contribution Award, which recognizes individuals who have made transformative contributions and advanced the frontiers of their field, akin to a lifetime achievement award in design. Second, the Annual Innovation Award, which rewards the design works that have a great impact and lead the future direction of design. Third,

问：能否谈谈您对本次大会主题——"设计无界，相融共生"的理解？

娄永琪：本次大会主题与"设计"概念的最新进展息息相关。回顾设计发展史，全球设计发展经历了若干阶段，基本每十年都会发生比较大的转折。

2015年，世界设计组织发布了最新版设计定义，指出设计最重要的三个贡献——第一是驱动创新，第二是推动商业成功，第三是创造更加美好的生活。设计如何完成这个使命？是通过设计创新型产品、创新型服务、创新型系统和创新性体验来完成的。

尽管这个定义已经是多重妥协的结果，但也还是可以看出，设计已经发生了巨大变化，远远超越了一般民众理解的建筑或产品造型范畴，从有形物的设计拓展到无形的服务、交互、体验乃至人工智能系统的设计。从设计发展角度看，设计疆域拓展和范式转型本身也是设计新陈代谢的进化需求。

而设计要发挥更大的作用，就要突破自己的专业圈层，更加勇敢地走到社会当中、走到产业当中、走到问题当中。既走出去，也要更融入、更交叉，从其他学科和行业汲取发展的动力，以激发更大范围的协同和共创现象发生，这就是所谓的"设计无界，相融共生"。

这事实上是一个全球趋势，并不仅仅是上海在做，而是由上海搭建了一个国际化的大舞台，邀请全世界的人结合各自面临的问题对这个设计议题进行思考，以此来形成上海答案。因此这个过程是开放的、动态的、交织的、碰撞的。我相信这个国际社群和其间的讨论会非常精彩，让我们一起期待吧。

the Future Innovation Award, which is the world's first design award targeting K12 students.

In addition, the Shanghai government is actively engaging various districts, universities, and organizations to host a diverse range of city-wide response activities. The four-day conference should be seen as just the beginning; the city-wide response and its ongoing influence will extend well beyond the event.

Q: Could you elaborate on your understanding of the theme of this conference —"Vision in Perspective: Design Beyond Borders, Diversity and Togetherness"?

Lou: The theme of this conference is closely linked to the latest developments in the concept of design. Looking back at the history of design, global design evolution has undergone several significant phases, with major shifts occurring approximately every decade.

In 2015, the World Design Organization released its latest definition of design, highlighting three key contributions: first, driving innovation; second, fostering commercial success; and third, creating a better quality of life. How does design fulfill this mission? By innovating products, services, systems, and experiences.

Although this definition represents a result of multiple compromises, it is evident that design has undergone substantial changes. It extends far beyond the general public's understanding of architecture or product aesthetics, evolving from tangible design to include intangible elements such as services, interactions, experiences, and even artificial intelligence systems. From the perspective of design development, this expansion of design domains and paradigm shifts are necessary for the evolution of design itself.

For design to make a greater impact, it must transcend its professional boundaries and engage more boldly with society, industry, and real-world issues. It must both reach out and integrate, drawing on insights from other disciplines and industries to stimulate broader collaboration and co-creation. This encapsulates the notion of "Vision in Perspective: Design Beyond Borders, Diversity and Togetherness".

This is indeed a global trend, not merely an initiative by Shanghai. Rather, Shanghai provides an international platform that invites people from around the world to address design issues in the context of their specific challenges, thereby contributing to Shanghai's response. Thus, this process is open, dynamic, intertwined, and confrontational. I believe that the international community and the discussions within it will be highly stimulating, and I look forward to it with great anticipation.

03 上海设计要勇当全国排头兵

问：设计人才培养在践行世界一流"设计之都"的建设中具有怎样的分量，现状如何？大会将如何推进设计人才队伍建设？

娄永琪：我认为，一座城市要建设成为世界设计之都，背后一定有一所或若干所顶尖的设计院校作支撑。你想，一座产出影响世界的设计思想、设计人才的城市，怎么可能不是世界设计之都？

设计人才培养很重要，但不能狭义化地考量。高等教育固然很重要，但如何推动上海更加多元的设计教育生态，更值得思考。对此，上海应致力于打造具有世界水平的多元设计教育生态，不仅包括一流综合性大学和独立设计学院，还要包括顶尖的职业教育、K12基础教育和终身学习教育。

此外，我们也要意识到，设计人才培养不仅发生在校园里，更发生在城市里。以同济大学"NICE2035未来生活原型社区"为例，我们通过一系列扎根社区的、"小而互联"的、跨学科整合和逆向创新的项目，推动大学知识和资源溢出，创造新模式和新经济，使社区从创新链和产业链的末端走向前端，成为城市创新的策源地。在这里，各种思想、知识、人群、创意、场景高度融合，社区就是没有围墙的大学。推而广之，上海整个城市就是一所没有围墙的创新大学，这是我对于未来上海设计之都建设的一个愿景。

目前上海的高等教育在培养设计人才方面的成绩是很不错的。以同济大学设计创意学院为例，在QS全球"艺术与设计"学科最新排名中位列第12名，连续五年亚洲第一，这既是同济大学学科建设的成果，也是上海高度重视设计人才培养收获的国际认可与反响。

近几年，通过上海设计学高峰学科平台，上海设计院校逐渐建立起同城协同高峰学科群。我相信在世界设计之都大会召开若干届后，除了同济大学之外，也会看到越来越多的上海设计院校进入世界设计学科排名前100的名单。我想，如果上海这座城市有3所各具特色的设计院校能进入世界前50阵营，那么上海建设世界一流"设计之都"的动力和潜力将更加永续。

除高等教育之外，上海在青少年设计人才培养中亦有所表现。在2020上海设计之都十周年系列活动上，

03 Shanghai Design Should Act as a Pioneer

Q: What is the role of design talent development in the pursuit of becoming a world-class "City of Design", and what is the current status? How will the conference advance the development of the design talent pool?

Lou: I believe that for a city to become a world-class City of Design, it must be supported by one or several leading design institutions. Consider a city that produces globally influential design ideas and talent—how could it not be a World Design City?

Design talent development is crucial, but it should not be considered in a narrow sense. While higher education is undoubtedly important, it is equally important to consider how to foster a more diverse design education ecosystem in Shanghai. Shanghai should aim to create a world-class, multifaceted design education environment that includes not only top-tier comprehensive universities and independent design schools but also leading vocational education, K12 education, and lifelong learning.

On the other hand, we must recognize that design talent development occurs not only in educational institutions but also within the city itself. For example, at Tongji University's project "NICE2035 Living Line", we are advancing a series of community-based, "small yet interconnected", interdisciplinary integration, and reverse innovation projects. These initiatives help to overflow university knowledge and resources, create new models and economies, and shift communities from the end to the forefront of the innovation and industry chain, making them sources of urban innovation. Here, various ideas, knowledge, people, creativity and scenarios are highly integrated, transforming the community into an university without walls. Extending this idea, the entire city of Shanghai could be viewed as an innovation university without walls. This is my vision for the educational space in Shanghai's future as a City of Design.

Currently, Shanghai has made impressive achievements in the higher education of design talent. For instance, College of Design and Innovation of Tongji University is ranked 12th globally in the latest QS World University Rankings for "Art and Design", maintaining the top position in Asia for five consecutive years. This is not only a testament to Tongji University's academic development but also an international recognition of Shanghai's commitment to design talent cultivation.

In recent years, through the Shanghai Design Discipline Peak Platform, design institutions in Shanghai have gradually established a collaborative peak discipline cluster within the city. I believe that after several editions of the World Design Cities Conference, we will see more Shanghai design institutions, in addition to Tongji University, entering the top 100 of the global

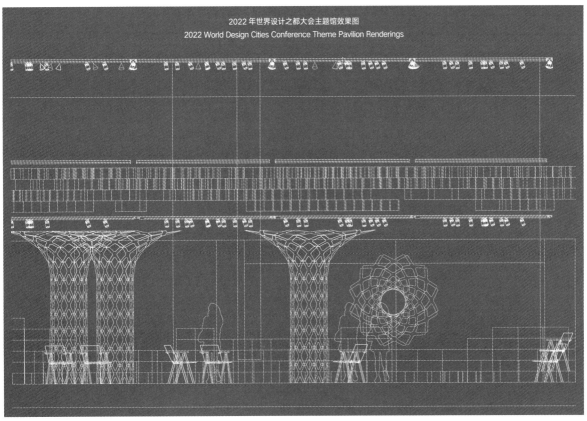

2022年世界设计之都大会主题馆效果图
2022 World Design Cities Conference Theme Pavilion Renderings

上海正式宣布成立上海市青少年创意设计院，上海市教育委员会邀请我担任该院首任院长，把设计带入上海所有中小学，培养未来设计师。值得一提的是，这是上海市教委在科艺融合的理念下，推出的第二平台院，第一平台院是上海市青少年科学研究院。由此可见，和前面讲的大会一样，上海在不同层面都有这样的思考：一手抓科技，一手抓创意，从这点讲，我对上海未来创意人才的培养特别有信心。

问：2022年是我国全面实施"十四五"规划的关键一年，立足这一时间节点，您认为未来上海创意设计产业的发力点与着力点应在何处？对此，您有着怎样的见解或建议？

娄永琪：记得几年前，《华尔街日报》曾发表一篇文章，说"设计学院正在成为未来的商学院"；《快公司》杂志也曾发文，"设计正在取代技术，成为硅谷创新的引擎"。我常在想，对于这些全世界在思考的问题和现象，上海能不能再往前走一步？这时候上海设计本身就应该像上海在全国的位置一样，要做排头兵，继续先行先试，开风气之先。

design discipline rankings. If Shanghai has three distinctive design schools ranked among the top 50 worldwide, the city's drive and potential for building a world-class "City of Design" will be even more sustainable.

Beyond higher education, Shanghai has also demonstrated progress in nurturing young design talent. During the "2020 Shanghai City of Design 10th Anniversary" series of events, Shanghai officially announced the establishment of the Shanghai Youth Creative Design Institute. The city's Education Commission invited me to serve as the institute's first director, aiming to introduce design into all primary and secondary schools in Shanghai and cultivate future designers. Notably, this institute is the second platform established by the Shanghai Education Commission under the concept of integrating science and art, following the Shanghai Youth Science Research Institute. This reflects Shanghai's broader approach to integrating technology and creativity, and I am particularly confident about the future cultivation of creative talent in Shanghai.

Q: 2022 is a crucial year for the full implementation of China's 14th Five-Year Plan, from this perspective, where should the future focus and efforts of Shanghai's creative design industry be directed? What insights or suggestions do you have on this?

2022年世界设计之都大会前沿设计创新奖
2022 World Design Cities Conference Frontier Design Innovation Award

首先，上海设计需要更加走向问题，要有更强的问题意识。哪里有问题，哪里就有设计。如果忘记这点，设计就容易闭门造车、自娱自乐。中国有很多独特的国情，社会发展中的很多新设计问题是之前其他国家没有遇到的，或者说我们设计的题材已经走到了世界的最前沿，比如乡村振兴、"双碳行动"、人工智能、元宇宙等，只有更贴近现实的真问题，我们才能产生伟大的设计思想和行动。

其次，上海设计需要更加走向产业。一是传统产业，通过设计提升发展能级，上海应该通过设计驱动，通过带动长三角区域级的产业转型升级；二是新兴产业，如新消费、大健康、新能源、新出行等，如何通过设计催生、壮大新的产业生态。产业的发展与更替日新月异，上海如何通过设计赋予未来更新的想象力，如何在战略性新兴产业赛道上走在世界最前列，这是上海需要思考的问题，也是上海的机遇。

我在一些演讲中发表过一个观点，未来最有钱的企业，或至少有一类企业，是"修地球"的。我们的地球已经无法承载人类这么大体量的活动，需要以产业化方式来"修复环境"，扩展环境容量。如果把"修地球"变成整个社会的基础产业，如同昔日的钢铁煤炭产业、今日的互联网产业一样，一旦成为基础产业、基础设施，就会有越来越多相关的新产业催生出来，从而助力上海"双碳"目标的早日实现。

最后，设计还需要进一步跨界。设计是人类一切有意识的创造活动的先导和准备，它不仅仅是一个专业，更是各行各业的人在思维底层与生俱来的一种素养，尽管这种素养往往因为社会的原因被深锁了，就好像孙悟空被困在五指山下一样。上海推崇设计，就是要把这张贴在"五指山"上的符揭掉，释放设计的魔力。如何团结尽可能多的人，如何让大家一起参与到设计中来，实现人人设计，设计事业才能发扬光大，这是上海设计要做的事，也是我们这次大会的一个长期目标。

补充一点，上海做任何事情，包括设计产业发展，都离不开全球性视野。我常问学生，上海"网红"建筑师邬达克是匈牙利建筑师还是上海建筑师？毋庸置疑，他是匈牙利人，但他几乎所有作品都留在上海，他职业生涯的绝大多数时间也在上海度过，从这个层面讲，他何尝

Lou: A few years ago, *The Wall Street Journal* published an article stating that "design schools are becoming the business schools of the future." Similarly, *Fast Company* magazine wrote that "design is replacing technology as the engine of Silicon Valley's innovation." I often wonder if Shanghai can advance even further in addressing these global issues and phenomena. In this context, Shanghai's design industry should position itself as a pioneer, just as Shanghai does nationally, by leading the way, continuing to experiment, and setting trends.

Firstly, Shanghai's design industry needs to become more problem-oriented and develop a stronger problem-awareness. Design arises where there are issues. If we lose sight of this, design risks becoming insular and self-indulgent. China has many unique national conditions, and many new design challenges we face are unprecedented globally. We are at the forefront of issues such as rural revitalization, carbon neutrality, artificial intelligence, and the metaverse, etc. Addressing real, pressing problems will lead to groundbreaking design ideas and actions.

Secondly, Shanghai's design industry needs to engage more with industry. Traditional industries can be elevated through design, with Shanghai leveraging design to drive the transformation and upgrading of industries in the Yangtze River Delta region. New industries, such as new consumption, health care, new energy, and new mobility, should also be nurtured and expanded through design. As industries evolve rapidly, Shanghai must use design to envision the future and lead in strategic emerging industries. This is both a challenge and an opportunity for Shanghai.

In some of my speeches, I have suggested that the wealthiest companies in the future, or at least a category of them, will be those focused on "repairing the Earth". Our planet can no longer support such an enormous scale of human activities, necessitating an industrial approach to environmental restoration and capacity expansion. If "repairing the Earth" becomes a foundational industry, akin to the steel and coal industries of the past or today's internet industry, it will generate new related industries, aiding Shanghai in achieving its carbon neutrality goals sooner.

Lastly, design needs to further cross disciplines. Design is the precursor and preparation for all human creative activities. It is not merely a profession but an inherent quality in the foundational thinking of individuals across various fields, although this quality is often suppressed by societal factors. By embracing design, Shanghai aims to remove these constraints and unleash the power of design. To truly advance design, we must engage as many people as possible and foster a culture where everyone participates in design. This is a long-term goal for Shanghai's design industry and for our conference.

不是地地道道的上海人?上海是一座海纳百川、兼容并包的城市,聚天下英才而用之,这是上海的气度。上海的设计产业发展也应如此,只有站在全球视野思考问题,才能体现上海的格局、上海的高度、上海的作为。

Additionally, Shanghai's design industry must operate with a global perspective. I have asked my students for many times whether the renowned architect László Hudec was a Hungarian or a Shanghai architect. Undoubtedly, he was Hungarian, but almost all his works were in Shanghai, and he spent the majority

of his career here. From this perspective, he could be considered a true Shanghai resident. Shanghai is a city that embraces all, welcoming and utilizing global talent. This spirit should also guide Shanghai's design industry, reflecting Shanghai's vision, stature, and achievements through a global lens.

第一部
Part One

2022年首届"设计无界,相融共生"
The 2022 Inaugural "Vision in Perspective: Design Beyond Borders, Diversity and Togetherness"

世界设计之都大会主题展——
World Design Cities Conference Theme Exhibition—

人·人,人·自然
People to People, Human to Nature

策展团队
Curatorial Team

娄永琪
Yongqi Lou

教授 同济大学副校长
英国皇家艺术学院荣誉博士
瑞典皇家工程科学院院士
Professor and Vice President of Tongji University
Honorary Doctor of the Royal College of Art
Fellow of Royal Swedish Academy of Engineering Sciences

娄永琪教授，博士，同济大学副校长，全国设计研究生教育指导委员会主任委员；长期致力于社会创新和可持续设计实践、教育和研究；在同济大学设计创意学院自创立（2009年）至2022年其卸任院长期间，推动学院成为世界一流设计学院；先后担任CUMULUS国际艺术设计院校联盟副主席、WDO世界设计组织执委、维也纳应用艺术大学国际咨询委员会主席。他是《设计、经济与创新学报》(爱思唯尔出版) 的创始执行主编，也是《设计问题》(麻省理工学院出版) 的编委，还是2021年国际设计研究大会联合主席，并受邀在2018年香港设计营商周、2017年国际室内建筑师 / 设计师团体联盟代表大会、2016年美国工业设计协会国际大会和2015年国际人机交互大会等著名会议担任主旨演讲人。他的作品和研究成果在芬兰赫尔辛基设计博物馆、意大利米兰三年展博物馆等处展出；他于2019年当选瑞典皇家工程科学院院士，于2023年获英国皇家艺术学院荣誉博士学位。

Prof./Dr. Yongqi Lou is a vice president of Tongji University and the Chair of the China National Graduate Education Steering Committee for Design. Lou has been at the forefront of the education, research, and practice of design for social innovation and sustainability. He has played a crucial role in the success of the College of Design and Innovation (D&I) of Tongji University, since its inception in 2009 and through his deanship till 2022. Lou has served on various boards, including CUMLUS, ICSID/WDO, DESIS, and Angewandte (Vienna). He is the founding executive editor of *She Ji: The Journal of Design, Innovation, and Economics* (published by Elsevier) and serves as an editorial board member of *Design Issues* (MIT Press). He was the co-chair of IASDR 2021 and was invited as a keynote speaker at prestigious events such as BODW 2018, IFI 2017, IDSA 2016, and ACM SIGCHI 2015. His works have been exhibited in the Design Museum of Helsinki, the Triennale Design Museum in Milan, and other renowned institutions. Lou was elected as a fellow of the Royal Swedish Academy of Engineering Sciences (IVA) in 2019, and in 2023, he received an honorary doctorate from the Royal College of Art.

奥雷·伯曼
Ole Bouman

国际设计协会理事,"设计互联"创馆馆长
Founder of Design Society
Board Member of Independent School for the City

陈伯康
Aric Chen

荷兰 Het Nieuwe Instituut 馆长、艺术总监
General and Artistic Director of Het Nieuwe Instituut

奥雷·伯曼,前深港城市/建筑双城双年展(2013—2014)创意总监,现任设计互联展览馆馆长,建筑学术杂志《体量》创始人兼主编。著有《看不见的建筑》(1994)、《责任建筑》(2009)。2006—2013年间,担任荷兰建筑协会(NAi)会长,2000年第三届欧洲当代艺术双年展Manifesta联合策展人,并为双城双年展(深圳)、圣保罗建筑双年展和威尼斯国际建筑双年展的策展作出重大贡献,曾在麻省理工学院教授建筑学。

Ole Bouman, former Creative Director of the Urbanism/Architecture Bi-City Biennale Shenzhen (2013-2014), is now the director of Design Society, Shenzhen. He was the founder and editor-in-chief of architecture magazine *Volume*. His publications include *The Invisible in Architecture* (1994) and *Architecture of Consequence* (2009). Between 2006-2013, Ole Bouman was the director of Netherlands Architecture Institute (NAi). He was co-curator of Manifesta 3 (2000), and was responsible for many entries to the architecture Biennales of Shenzhen, São Paulo and Venice. Ole Bouman taught architecture at the Massachusetts Institute of Technology.

陈伯康,荷兰Het Nieuwe Instituut馆长兼艺术总监。该馆是位于荷兰鹿特丹的国立机构,亦是一所有关建筑、设计与数字文化的主题博物馆。生于美国的陈伯康曾出任同济大学设计创意学院策展实验室主任、美国设计迈阿密(Design Miami)策展总监及北京设计周的创意总监。此外,他亦是香港M+博物馆设计与建筑领域的首席策展人,负责主理博物馆有关设计和建筑的馆藏和项目发展。

Aric Chen is General and Artistic Director of Het Nieuwe Instituut, the Dutch national institute and museum for architecture, design and digital culture in Rotterdam. American-born, Chen previously served as Professor and founding Director of the Curatorial Lab at the College of Design and Innovation of Tongji University in Shanghai; Curatorial Director of the Design Miami fairs in Miami Beach and Basel; Creative Director of Beijing Design Week; and Lead Curator for Design and Architecture at M+ in Hong Kong, where he oversaw the formation of that new museum's design and architecture collection and program.

潘爱心
Azinta Plantenga

策展实验室副主任 同济大学设计创意学院
Associate Director Curatorial Lab
College of Design and Innovation, Tongji University

魏天天
Blair Wei

瑞安新天地生态合作及创新投资负责人
Head of Ecosystem & Venture Investment
Shui On Xintiandi

潘爱心在过去五年中一直在中国从事策展人、设计历史学家和研究人员的工作。她目前是同济大学设计创意学院的助理研究员，教授设计历史与批评，并担任同济大学设计创意学院策展实验室的副主任。此前，她于2017—2019年在深圳设计互联担任策展人。2017年前，她在阿姆斯特丹市立博物馆的设计和装饰艺术策展团队工作。她来自荷兰，拥有莱顿大学的中国研究和设计史学位。

魏天天，毕业于哈佛大学建筑系及康奈尔大学建筑系。负责XINTIANDI新天地在文化消费及艺术创意领域的战略规划及创新投资，包括"新天地"与UCCA共创的"燃冉"青年艺术家孵化计划在内的重点文化战略项目，持续扶持具有在地精神的文化艺术生态。加入瑞安集团前，在上海从事私募股权投资工作。她也曾就职于OMA及SHoP Architects，亦参与了第15届威尼斯建筑双年展"共享•再生"平行展的策展工作。

Azinta Plantenga has been working as a curator, design historian and researcher in China for the past 5 years. She is currently Assistant Research Fellow at College of Design and Innovation of Tongji University in Shanghai, where she teaches Design History and Criticism and is Associate Director of the Tongji D&I Curatorial Lab. Previously, she was a curator at Design Society in Shenzhen from its establishment in 2017 until 2019. Before moving to China she worked for the Stedelijk Museum Amsterdam (NL) Design and Decorative Arts curatorial team. Originally from the Netherlands, she holds a degree in Chinese Studies and Design History from Leiden University.

Blair owns a B.Arch degree from Cornell University and an M.Arch degree from Harvard Graduate School of Design. Blair is currently in charging of cultural strategy planning, partnership building and venture investment for the XINTIANDI community, playing an leading role in the XINTIANDI x UCCA young artist incubation program "RanRan". Before joining Shui On, she was a real estate private equity investment professional. Blair has also practiced at OMA and SHoP Architects before returning to Shanghai and worked as part of the curatorial team for the 15th Venice Biennale Themed Exhibition "Sharing & Regeneration" with Fondazione EMGdotART.

助理策展人 刘畅、陶思旻、王依琳、黄粒莹、纪丹雯、刘曌
空间及平面设计 郭泠、杜钦、丁卓媛、尤优

Assistant Curators Chang Liu, Simin Tao, Yilin Wang, Liying Huang, Danwen Ji, Zhao Liu
Spatial and Graphic Design Ling Guo, Qin Du, Zhuoyuan Ding, You You

策展序言
Curatorial Statement

本次展览聚焦一个基本而紧迫的问题，即人们如何能够在彼此之间、与自然之间，以及与我们生活的星球之间创造新型、可持续、有韧性、充满活力的关系。

我们关注已有的种种努力和成就，关注人们如何摒弃短视思维，利用创造力、信任和乐观精神打造新经济、新文化和新型平衡关系，共同拥抱一个更加可持续的未来。在此过程中，设计行业正加速转型，实践方向从推出新颖独特的产品，转变为利用设计领域的综合技能——从感知和构想，到原型制造和规模扩增，来应对基本而紧迫的挑战。

展览围绕生命中的六大挑战展开叙事，前五个主题对应的这些挑战均源于人类的基本需求或生存驱动力："食物""健康""信任""环境"和"技术"。而主题6，即结语部分的"繁荣"既是挑战，又是人类文明的成就，颂扬生命的丰富多彩，并揭示出一个不断壮大、充满希望、主动担当的创意社群所具有的重要意义。我们相信，上海世界一流设计之都的建设，将为这个社群的成长提供丰富的营养；与此同时，这个社群也将为上海建设世界一流国际化大都市增添创意的动能。

This exhibition addresses the fundamental and urgent question of how people can create new, sustainable, resilient, and vibrant relationships with each other, with nature and the planet we live on.

We focus on the efforts and creations that are taking place, on how people abandon short-sighted thinking and use creativity, trust, and optimism to design new economies, new cultures, and new balanced relationships, embracing a more sustainable future together. In the process, the design field is accelerating its shift from a practice of proposing new unique products, to a practice of responding to essential and urgent challenges with design's comprehensive skill set, ranging from sensing and envisioning, to prototyping and scaling.

The narratives of this exhibition are to be told by presenting six life challenges, all stemming from fundamental human needs or drives: food, health, trust, environment, technology, and the Theme 6 "prosperity", as the conclusive one, indicates not only the challenge but also the achievement of human being, honoring a full spectrum of life, and revealing the purpose of a growing, hopeful, proactive, and creative community. We believe that the development of Shanghai as a world-class city of design provides rich growth nutrients for such community. In turn, this group of people adds creative momentum for the development of Shanghai as a world-class international metropolis.

策展结构：六大叙事
Curatorial Structure: Six Major Narratives

主题展的叙事将从六大挑战来讲述,这些挑战都源于基本需求:食物、健康、信任、环境、技术与繁荣。从生活必需品到文明的愿望,它们都属于生活的范畴。

主题展展示了设计实践向全面解决这些问题在言行上的转变,包括个人与集体,成熟的作品与新兴的声音。

The narratives of the exhibition will be told by presenting six challenges, all stemming from fundamental needs: food, health, trust, environment, new technologies and prosperity, ranging from life essentials to civilizational aspirations. They all belong to the spectrum of life.

The exhibition demonstrates the shifting practice of design to a comprehensive resolve of these issues, in words and in deeds, by individuals and collectives, by mature work and emerging voices.

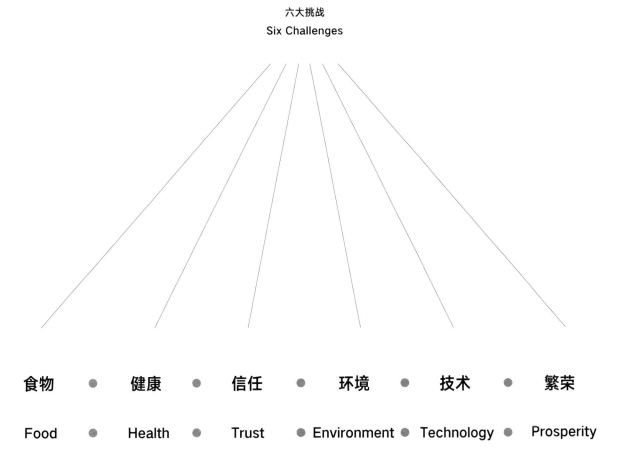

作品介绍
Collections

主题1: 食物
Theme 1: Food

稳定的食物设计如何展现一种创新文化？

没有食物和洁净的水，就没有我们人类。人生在世，离不开每日膳食。但如今，这个对于恢复体力的永恒不变的条件正面临剧变。一方面，是由于食物的基本内容正在发生变化，包括食物成分、营养物质、饮食方式及膳食模式；另一方面，是由于经过多年工业化和大规模生产，食品生产领域的创新层出不穷，体现在种植模式、耕作技术、收获和包装、分销和营销等各个方面。倘若不改变对待食物的态度，我们将无法实现生态平衡。但这种改变本身又将帮助我们重新定义这一生态，并从中受益。与食物安全相关的危机存在于其生产、加工、流通、储存、分配、消费等各个环节。以上任何一个环节出了问题，人类就会遭受巨大伤害。这些都是设计应该努力的方向。

从一粒米可以窥见中国几千年的历史。反之，粮食在21世纪的发展历程，也映衬出中国的发展历程。好的食物助力铺就和谐之路。

How does stable food design represent a culture of innovation?

We are nothing without food and clean water. In one's lifetime, one needs daily meals. But this timeless condition of physical recuperation is in great upheaval today. Partly, because food's very substance is changing: its constituents, its nutrients, its diets, and its menu. Partly, because the production of food, after years of industrialization and mass production, is full of innovation: in modes of cultivation, in farming techniques, in harvesting and packaging, distribution and marketing. We cannot establish a balanced ecology without a changing attitude towards food. But this change will in itself help us re-define this ecology and benefit from it. The crisis related to food safety exists in various aspects such as production, processing, circulation, storage, distribution, and consumption. If any link goes wrong, humanity will suffer huge harm. These are all directions that design should strive towards.

For thousands of years, the Chinese world could be found in a grain of rice. In reverse, the 21st century adventures of such grain, can be the adventure of China. Good food is the gateway to peace.

正阳定食
Zhengyang Food Set

上海农场 | 2019 年至今
Shanghai Farm | 2019 till now

民以食为天。中国拥有庞大的人口基数，如何保障从农场到餐桌的安全和新鲜的食品供应，并改善市民的饮食结构，是一个复杂的系统设计问题。隶属于光明食品上海农场的"正阳定食"，是一家面向学校、政府、企业，提供标准的生鲜半成品和冷热膳食的现代化中央厨房企业。"正阳定食"通过智能技术赋能的"中央厨房+乡村振兴"模式，以供应学生营养餐为切入点，全面构建起"从农场到餐桌"的全产业链，实现了农场功能拓展和城市餐饮供给保障的多目标协同。

Food is essential to life. With China's vast population, ensuring the safety and freshness of food from farm to table while also improving urban dietary structures presents a complex design challenge. "Zhengyang Food Set" affiliated with Guangming Food's Shanghai Farm, is a modern central kitchen enterprise that provides standardized fresh semi-finished products, and hot and cold meals to schools, government bodies, and businesses. By leveraging smart technology in its "Central Kitchen and Rural Revitalization" model, Zhengyang Food Set focuses on providing nutritious meals for students. It has built a comprehensive farm-to-table supply chain, achieving multiple goals in expanding farm functions and ensuring a stable urban food supply.

安和托皮亚屋顶温室
Rooftop Greenhouse Agrotopia

van Bergen Kolpa 建筑事务所，META 建筑事务所 | 2021
van Bergen Kolpa Architects, META Architectuurbureau | 2021

都市农业是集生产、教育、交流及体验于一身的创新农业形态。安和托皮亚是欧洲规模最大的都市农业生产和研究中心之一，也是一座向大众开放的公共建筑。温室采用大面积双层玻璃幕墙、纪念碑式入口楼梯和层叠功能体块，形成了引人注目的外观。通过集约式空间、雨水收集、循环用水、利用周边垃圾焚烧设施的市政余热供暖等可持续技术，温室成为整合美观和绿色功能的风向标。温室周围环绕一条公众参观路线，加强了高科技智慧设施农业和公众的互动，实现了与城市循环共生的关系。由此，市民不仅可以学习种植，还能参与到新园艺技术和商业模式开发中。

Urban agriculture is an innovative form of agriculture that combines production, education, communication, and experience. Agrotopia, one of the largest urban agriculture production and research centers in Europe, is also a public building open to visitors. The greenhouse features expansive double-layer glass curtain walls, monumental entry stairs, and cascading functional blocks, creating a visually striking structure. Utilizing sustainable design technologies such as efficient space use, rainwater collection, water recycling, and municipal waste heat recovery from nearby incineration facilities, the greenhouse sets a benchmark for blending aesthetics with green functionality. A public visitor route encircles the greenhouse, enhancing interaction between the high-tech facilities and the public, and fostering a symbiotic relationship with the urban environment. Visitors can not only learn about planting but also participate in the development of new horticultural technologies and business models.

兜着走
FooDZZ

咪咪 | 上海夕食科技有限公司 | 2022
Mimi | Shanghai Xishi Technology Co., Ltd. | 2022

食物短缺的问题的解决，一方面需要"开源"，也就是提升粮食的产能；另一方面，也需要"节流"，也就是尽可能减少食物的不必要浪费。"兜着走"是一个关注食物浪费问题的资源整合平台，鼓励向善的消费。"兜着走"本身是一个非常典型的产品服务体系设计。平台通过微信小程序发布折扣商品和"剩食盲盒"信息，为商家提供更可持续的解决方案，让消费者在买到低价安全食物的同时，减少食物浪费。合作商家包括面包店、超市及各大食品品牌，折扣商品涵盖卖不完的面包和临近质保期的食物等。此外，"兜着走"还提供食物储存手册和食用指南，引导消费者树立正确的剩食观念，从消费端减少浪费。

Addressing food shortages requires both "increasing supply"—boosting food production—and "reducing waste"—minimizing unnecessary food loss. FooDZZ is a resource integration platform focused on food waste issues, promoting responsible consumption. It is a prime example of product-service system design. The platform uses a WeChat mini-program to share information on discounted items and "surplus food mystery boxes" providing merchants with more sustainable solutions while enabling consumers to purchase affordable, safe food and reduce waste. Partner merchants include bakeries, supermarkets, and major food brands, offering discounts on unsold bread and items nearing their expiration dates. Additionally, FooDZZ provides food storage manuals and consumption guides to help consumers adopt better practices for managing surplus food and reducing waste.

蟲托邦
Insectopia

张之弦，周雯萱 | 好公社 | 2021
Zhixian Zhang, Wenxuan Zhou | NICE COMMUNE | 2021

可食用昆虫是一种新型食源，富含蛋白质、脂肪、碳水化合物等有机物质，及钾、钠、磷、铁、钙等无机物质，还有人体所需的游离氨基酸。在不少地方，昆虫原料已成为食品行业的一个新兴趋势。但如何破除人们充分利用这种食源的心理障碍是一个综合的设计挑战。同济大学设计创意学院学生张之弦和周雯萱在2021年发起了蟲托邦（Insectopia）项目，结合"insect"（昆虫）和"utopia"（乌托邦），旨在为消费者提供更具吸引力的昆虫消费场景。

蟲托邦关注"吃虫"以及可持续、食物、自然、艺术等议题。

Edible insects are an emerging food source, rich in proteins, fats, carbohydrates, and essential nutrients such as potassium, sodium, phosphorus, iron, calcium, and the free amino acids required by the human body. In many regions, insect-based ingredients have become a growing trend in the food industry. However, overcoming the psychological barriers to fully embracing this food source poses a significant design challenge. In 2021, Zhixian Zhang and Wenxuan Zhou, students from College of Design and Innovation of Tongji University, launched the project "Insectopia", merging the concepts of "insect" and "utopia" to create a more appealing context for consuming insects.

"Insectopia" explores themes of insect consumption, sustainability, food, nature, and art, aiming to provide consumers with an engaging and thought-provoking experience with edible insects.

韧性社区的食物系统研究与设计
Research and Design of Food System in Resilient Community

梁茹茹 | 清华大学美术学院 | 2022
Ruru Liang | Academy of Arts & Design, Tsinghua University | 2022

如何将都市农业融入城市社区？该设计是一套基于韧性社区理念的在地化食物生产系统：它结合了自然农业和受控环境农业的优势，利用无土栽培技术，在有效提升社区食物系统风险应对能力的同时，还提供居民体验种植和交流互动的空间，并且具有社区景观及食物科普的功能。因此，该设计结合社区公共空间种植蔬菜等作物的全新解决策略，为重新定义农业和市民之间的交互提供了新的可能。

How can urban agriculture be integrated into city communities? A new strategy that growing vegetables and other crops in public community spaces offers innovative ways to redefine the interaction between agriculture and residents. This design presents a localized food production system based on the concept of resilient communities. By combining the benefits of natural farming and controlled environment agriculture, and utilizing soilless cultivation techniques, the system not only strengthens the community's ability to manage food system risks but also provides spaces for residents to participate in gardening and social interaction. Additionally, it serves as a community landscape feature and a platform for food education.

食物的循环设计
Circular Design for Food

艾伦·麦克阿瑟基金会 | 2021
Ellen MacArthur Foundation | 2021

循环经济通过资源和产业循环取代线性经济,以减少废弃物产生、延长产品和材料的使用周期、促进自然资源再生,以更好地应对气候变化、生物多样性丧失等全球性挑战。艾伦·麦克阿瑟基金会是全球循环经济的重要倡导者之一,其报告《重塑食物:利用循环经济促进自然再生》展示了食品行业如何通过重新设计产品,影响什么食物被食用、什么原料被生产及其生产方式,并与阿尔法食物实验室合作,运用思辨设计的方法,畅想了四款在2030年的超市货架上随处可见的自然友好产品。

The circular economy replaces linear models with cycles of resources and industries to reduce waste, extend the lifecycle of products and materials, and promote the regeneration of natural resources. This approach is aimed at more effectively addressing global challenges such as climate change and biodiversity loss. Through its report *The big food redesign: regenerating nature with the circular economy*, the Ellen MacArthur Foundation, a leading advocate for the circular economy, demonstrates that circular design for food offers FMCGs and retailers a pathway to realise food's potential to be good for nature, farmers, and business. Working with Alpha Food Labs, they used speculative design to imagine four nature-friendly products that will be on supermarket shelves in 2030.

多样化的原料

为促进动植物的遗传多样性,从而建立有韧性的食物供应,各企业可将更多样的原料引入其产品组合。例如,在烹饪时所需的甜味不仅可以提取自甘蔗、甜菜或玉米,也可从多年生作物如椰枣、角豆和椰子以及高甜度天然甜味剂罗汉果和甜叶菊中提取。该原则适用于各类原料。比如种植一系列不同品种的作物,例如小麦,可提高全球小麦生产抵御外部冲击的能力。

低影响的原料

从常规生产的动物产品转向低影响的替代品,从高环境影响的作物转向低影响的作物是立竿见影的举措。许多企业已经在探索从以常规方式生产的动物蛋白转向植物蛋白的可能性。本研究表明,这类机遇远远不限于拓展蛋白质的来源。例如,在被调研地区域内,将一盒早餐麦片中的常规小麦粉换成豌豆粉可减少原料种植农场40%的温室气体排放,将该农场对生物多样性的负面影响降低5%。

升级利用的原料

当前,高达三分之一的食物被损失或浪费。升级利用方面的创新不仅可以避免食物和副产品进入垃圾填埋场,更可以将它们转换成高价值原料。在新技术的推动下,当前市值460亿美元的升级利用食品市场预计将以每年5%的速度增长。快消品企业和零售商可以通过推广该类解决方案,挖掘不断增长的市场机遇。使用升级利用的原料还可减轻土地压力,实现土地、能源和其他用于种植食物的投入回报最大化。

以再生农业方式生产的原料

近年来,领先企业已经认识到再生农业生产的环境效益。除了更高的产量,再生农业生产还能显著提高农民的经济效益。没有方法可放之四海而皆准,因此需要时间反复摸索和验证。不过,对于本报告中所有被建模研究的原料来说,因地制宜的实践方法可在向再生农业的转型期结束后提高粮食总产量,为农民提供额外收入,同时产生显著的气候和生物多样性效益。

渣渣简约休闲椅 / 哲思直饮杯
Kafftec Chair/ Kafftec Cup

咖法 | 2022
Kafftec | 2022

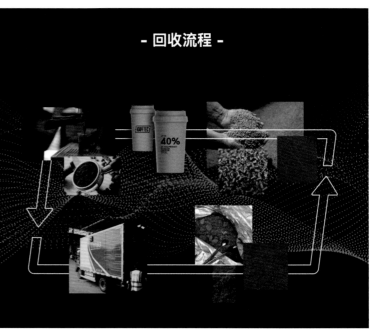

"人们喝咖啡的时候，忽然意识到只有0.2%的内容进入了口中，剩下的99.8%则是以咖啡渣的形式被丢弃了。他们会想，'我做了些什么啊！'或者应该这么说：'我有哪些事没做呢？'"《蓝色经济》一书的作者冈特·鲍利对咖啡渣再利用情有独钟。他将咖啡渣变成肥料、种植蘑菇，甚至将之改造为布料、颜料及发电原料，以此创造就业，带动经济。无独有偶，咖法是一个专注咖啡渣回收并循环再造全新产品的创新环保品牌，提供回收升级的材料和产品及环保解决方案，助力品牌和企业实现循环经济，为可持续生活方式提供更多选择。目前咖法已拥有35项国家专利，9项国际专利，材料和产品符合FDA和ROHS指令标准。量产商品的塑料替代率最高可达60%，并多次获得国内外知名设计奖项。

"When people drink coffee and they suddenly realize that it's only 0.2% that goes into their mouth and 99.8% is wasted as coffee ground. They feel like: 'Oh, what have I been doing!' Or, should the question be instead: 'What have I not been doing?'" Gunter Pauli, the author of *The Blue Economy*, is passionate about reusing coffee grounds. He has turned them into fertilizers for mushrooms growing, or even transformed them into fabrics, dyes, and energy sources, creating jobs and stimulating the economy. Similarly, Kafftec is an innovative and eco-friendly brand focused on recycling coffee grounds to create new products. Kafftec offers upgraded materials and products, providing sustainable solutions that help brands and businesses achieve a circular economy while offering more options for sustainable living. Currently, Kafftec holds 35 national and 9 international patents, with its materials and products meeting FDA and ROHS standards. The plastic replacement rate for mass-produced items can reach up to 60%, and Kafftec has received numerous prestigious design awards both domestically and internationally.

为农业系统而设计：中国农谷·农创中心

Design for the Agricultural System: Agricultural Innovation Center, China Agricultural Valley

左靖工作室 × typo_d 工作室 | 2022

Zuo Jing Studio × typo_d | 2022

2022年5月，为了荆门地方农业系统的整体性创新发展，左靖工作室策划设计了"中国农谷·农创中心"。设计团队首次使用了"地方设计"的概念，旨在强调乡村社会设计与具体地方的关系。作为一个农创活动体，农创中心将农产品区域公用品牌的视觉提升、地方农耕文化的挖掘展示，以及食农教育与体验融为一体，是将农业与文化、艺术与设计、食物与赏味相结合的综合服务场所。其中，对覆盖全市域、全品类、全产业链的农产品区域公用品牌"荆品名门"的打造，是此次为农业系统设计的一大重点。由typo_d工作室负责的品牌VI以及产品包装的全面升级，助力城市消费者与农产品生产者的链接。

In May 2022, the Agricultural Innovation Center (China Agricultural Valley), devised by Zuo Jing Studio to benefit the overall pioneering development of the local agricultural system in Jingmen, was officially unveiled. This is the studio's first use of the concept of "local design" to emphasize the relationships between rural social design and specific places. As an innovative agricultural complex, Agricultural Innovation Center integrates the visual enhancement of regional public brand of agricultural products, the excavation and display of local agrarian culture, and the education and experience of food and agriculture. It offers a comprehensive service that combines agriculture and culture, art and design, as well as food and taste. The design of Jingmen Regional Public Brand Association of Agricultural Products, a brand that covers all categories and industrial chains of the entire city, is a major focus of this project. The comprehensive upgrade of brand VI and product packaging, in charge by typo_d studio, enables the Jingmen Regional Public Brand to use innovative visual communication to link urban consumers and agricultural producers.

主题2: 健康
Theme 2: Health

大众交通方式如何展现健康出行的创新文化？

How does a popular mode of transport represent a culture of innovation to healthy mobility?

健康似乎是一种中性状态，人们要么身体健康，要么遭受各种疾病困扰，失去健康。健康似乎也是个人私事，人们要么通过健康生活方式保持健康，要么遭遇厄运，或者由于不良生活方式而引起疾病甚至死亡。

然而，历史表明，健康在很大程度上依赖于周围的环境，取决于诸多因素，远非个人能力可以控制。健康取决于人们的居住环境，取决于父母的健康状况，取决于科学进步，取决于技术能力，也许最重要的是，取决于人们的信念和心态。当前的新冠疫情再次充分说明，不同的思维方式，以及对长期健康福祉与短期利益优先性的不同考量会产生巨大的影响，并对整个人类文化带来持久的影响。健康问题涉及生活标准、生活方式、环境、社会政策和医疗服务等方方面面，所有这些影响因素错综复杂、盘根错节。请骑上自行车去探索一番，了解从与健康相关的产品、服务、沟通、交互、体验、环境、系统到机制，设计能够为此带来哪些改变。

Health may seem like a neutral thing, you either are healthy, or you are not, suffering from all kinds of diseases. It may also seem to be a personal thing, you either stay healthy by healthy living, or you run into bad luck, or your lifestyle inflicts sickness and premature death.

However, history shows how health is highly connected to the surrounding environment and depends on numerous factors, beyond one's reach. It depends on where you live, on the health of your parents, the scientific progress, the technological set up, and perhaps most of all, on people's beliefs and mind set. The current pandemic is, once again, a perfect example of how different ways of thinking, and different priorities of health for the long term versus profit for the short term, can have enormous consequences, and mark entire cultures. Health is a battlefield on which life standards, lifestyle, environment, social policies, and medical care collide. Please step on your bicycle and explore where design can make a difference to health-related products, services, communication, interaction, experience, environment, systems, and mechanisms.

"竹马(笃)"竹自行车
"Bamboo Horse" Bicycle

杨文庆 | 龙域设计 | 2013
Wenqing Yang | LOE Design | 2013

本产品通过可持续设计助力老字号创新，倡导绿色出行，解决最后一公里问题。"竹马"是一款符合工业化量产的竹自行车，采用竹钢混合结构，竹材坚韧轻质、快生环保。竹子具有高度的韧性和良好的吸震特性，相比普通钢管自行车可减重 30%~40%。经过加工的竹子能够承受上万次振动和疲劳测试。创新钢竹嵌合工艺发明专利，使自然生长的竹子和标准化钢材相匹配，既符合量产需求，又保留了传统人文特质。竹马不仅仅是一辆自行车，更是一种生活方式的体现。竹马代表的，是人们翘首以盼在个人交通领域产生的一种理念上的升级，也是一种结合节制与经典的新方式。

This product exemplifies sustainable design by innovating within a traditional brand and advocating for green transportation solutions. Designed to address the last-mile problem, the "Bamboo Horse" bicycle is an industrially mass-produced bamboo bike featuring a bamboo-steel hybrid structure. Bamboo is renowned for its toughness, light weight, and rapid growth, making it an environmentally friendly choice. Its high resilience and excellent shock absorption properties result in a 30%~40% weight reduction compared to traditional steel bikes. The processed bamboo can withstand thousands of vibrations and fatigue tests. The innovative aluminum-bamboo joint technology, patented for its invention, harmonizes naturally grown bamboo with standardized steel, meeting production demands while preserving traditional cultural elements. The product represents more than just a mode of transportation; it embodies a conceptual upgrade in personal transport, blending restraint with classic design.

共呼吸
Co-breath

孙苡茗 | 同济大学设计创意学院 | 2022
Yiming Sun | College of Design and Innovation, Tongji University | 2022

人们越来越渴望有一个健康的呼吸环境。作为实验性、自发式提供呼吸健康的穿戴设备，共呼吸利用小球藻的高光合能力，将人体呼吸产生的二氧化碳转化为养料，释放氧气，形成自驱动的气体循环系统。通过人类呼吸来喂养小球藻，并利用自然光合作用将二氧化碳循环为新鲜氧气，共呼吸成为一个自我维持的生态系统。共呼吸可根据用户活动水平和个人需求定制氧气流量，提供个性化的体验。凭借多种产品原型，共呼吸满足了市场对空气污染解决方案和对健康生活方式的需求，具备广泛的商业应用潜力。

There is an increasing demand for a healthy breathing environment. Co-breath is an experimental device that provides respiratory health. It utilizes microalgae's photosynthesis to convert the carbon dioxide from human exhalation into oxygen, forming a self-sustaining system. By using human breath to feed microalgae and harnessing photosynthesis to recycle carbon dioxide into oxygen, Co-breath could provide environmental and health benefits. Co-breath customized oxygen flow based on activity levels and personal needs, providing a personalized experience. With multiple product prototypes, Co-breath has commercial potential in everyday life. This enables various product prototypes for commercial use and tap into the growing market for air pollution solutions and wellness.

曹杨百禧公园
Caoyang Centennial Park

刘宇扬 | 刘宇扬建筑事务所 | 2020—2021
Yuyang Liu | Atelier Liu Yuyang Architects | 2020-2021

摄影：朱润资

城市迅速发展所导致的邻里环境品质下降等问题，使人们追求良好"人居环境"的积极性日益高涨。曹杨百禧公园项目基地长近1km，宽度介于10m至15m之间，前身为真如货运铁路支线，后改为曹杨铁路农贸综合市场。2019年，该市场关停后，这个空间在不到一年的时间内被重新规划建设为一个全新的、多层级、复合型步行体验式社区公园绿地。百禧公园以"3K"通廊为概念将艺术融入曹杨社区生活，从多维度回应2021年上海城市空间艺术季。设计通过挖掘场地文脉、建构空间场景，得以重塑街道绿网，形成"长藤结瓜"般的南北贯穿的步行纽带，进一步拓展曹杨社区的有机更新。

Rapid urban development has led to a decline in neighborhood quality, increasing the demand for a better "living environment". The Caoyang Centennial Park project spans nearly 1 kilometer in length and varies between 10 to 15 meters in width. Originally a freight railway branch, it was later converted into the Caoyang Railway Agricultural Market. After the market's closure in 2019, the space was redesigned and constructed in less than a year into a new multi-layered, mixed-use pedestrian community park. Conceptualized around the "3K" corridor, the park integrates art into the Caoyang community, responding to the 2021 Shanghai Urban Space Art Season from multiple perspectives. The design, which explores the site's historical context and constructs spatial scenarios, revitalizes the green street network, creating a fruiting-vine-shaped north-south pedestrian link and further extending the organic renewal of the Caoyang community.

可移动式"火眼"实验室系列产品
Movable "Huo-Yan" Product Series

上海易托邦建筑科技有限公司 | 中国设计智造大奖金奖作品 | 2021
Shanghai Etopia Building Technology Co., Ltd. | DIA GOLD | 2021

本系列产品在新冠疫情期间由同济大学设计创意学院与华大基金联合开发。城市面对突如其来的疫情挑战，其防疫基础设施储备往往显得力不从心。创新性的应急产品设计可以快速响应这些需求的挑战。可移动式"火眼"系列产品包含核酸检测实验室和隔离病房，可通过空运精准投放至全球各地。"火眼"的设备供应链也从工业级转向民用级，并采用分布式生产的方法，提高生产速度、降低量产成本。得益于"火眼"系统的创新设计解决策略，它可以低成本地快速提升一个城市的核酸检测能力，因此在新冠疫情期间大显身手，在全世界8个国家和数十个城市得到应用。

Developed collaboratively by College of Design and Innovation of Tongji University and the BGI Foundation during the COVID-19 pandemic, this product series addresses the urgent need for rapid response in pandemic situations. Traditional urban infrastructure often struggles to meet sudden challenges, underscoring the need for innovative emergency designs. The movable "Huo-Yan" series includes nucleic acid testing laboratories and isolation wards. These units can be transported in inflatable and vehicle-mounted containers, enabling precise air transport and global deployment. The "Huo-Yan" supply chain has transferred from industrial-grade to civilian-grade and utilizes distributed manufacturing to enhance production speed and reduce costs. Thanks to its innovative design, the "Huo-Yan" system has significantly increased nucleic acid testing capabilities at a low cost during the pandemic and been deployed in eight countries and dozens of cities worldwide.

分布式的制水净水设备研究与设计：以金平县农村为例
Research and Design of Distributed Water Preparation and Purification Equipment — Taking Rural Areas in Jinping City as an Example

黎书 | 清华大学美术学院 | 2022

Shu Li | Academy of Arts & Design, Tsinghua University | 2022

Calliope-Data：面向大众的可视数据故事智能设计生成系统
Calliope-Data: An Intelligent Visual Data Story Generation System for Mass

曹楠，石洋，陈晴，史丹青，吴妍秋，许欣悦，孙馥玲，郭熠，陈楠，孙梦迪，蓝星宇，刘佩，陈思极，彩丽甘，刘恣嫣 |
同济大学设计创意学院 iDVX 实验室 | 2023

Nan Cao, Yang Shi, Qing Chen, Danqing Shi, Yanqiu Wu, Xinyue Xu, Fuling Sun, Yi Guo, Nan Chen, Mengdi Sun, Xingyu Lan, Pei Liu, Siji Chen, Ligan Cai, Ziyan Liu |
iDVX Lab, College of Design and Innovation, Tongji University | 2023

本设计基于分布式系统设计理论，针对云南省金平县农村缺水的情况，设计研发了分布式水净化系统，同时开发了系列制水净水设备。净水设备应用了石墨烯材料，可以显著提高制水与净水的效率，提升当地供水系统的灵活性和多样性。

Calliope-Data是一款可视数据故事智能设计生成系统，能帮助用户了解数据中蕴含的关键信息，并将数据自动转化为有价值的多视图数据可视化报表，从而为用户的决策过程提供内容翔实且生动形象的数据参考。该系统提供了便捷高效的编辑功能，通过一系列交互手段，用户能够方便地修改系统生成的可视化报表及数据故事，以一种便捷友好的方式完成复杂的数据任务。它为用户提供了快速、便捷、直观、智能化的数据服务，降低了技术门槛，提供了全新的用户体验。

Based on the theory of distributed system design, this project addresses water scarcity in rural areas of Jinping County, Yunnan Province. It involves the development of a distributed water purification system and a series of water preparation and purification devices. The equipment utilizes graphene materials, which significantly enhance the efficiency of water preparation and purification while improving the flexibility and diversity of local water supply systems.

Calliope-Data is an intelligent visual data story generation system that helps users understand the critical information contained in their data. It automatically transforms data into valuable, multi-view data visualization reports, providing detailed and vivid data references for decision-making process. The system provides convenient and efficient editing features. Through a series of interactive operations, users can easily modify the visualization reports and data stories generated by the system. It provides users with fast, convenient, and intuitive data services. It greatly lowers the technical barriers and provides a different user experience.

农村蹲坐一体式无水便器及旱厕系统设计
An Integrated Dual Use Toilet and Waterless Disposal System for Rural Areas

刘新,严泽腾,梁骥,武洲 | 清华大学美术学院 | 2019

Xin Liu, Zeteng Yan, Ji Liang, Zhou Wu | Academy of Arts & Design, Tsinghua University | 2019

针对中国农村地区有老人的家庭设计的免水冲厕所,包括蹲、坐两种不同模式,并配有粪尿收集以及资源化利用系统。作品立足于中国农村现状,设计出一套人性化的、具有中国特色的过渡型便器产品。该设计既保留了传统蹲便式如厕习惯,又考虑到老年人对坐便器不断增长的需求;产品经过严格的人机工学测试与评估,采用EPP弹性泡沫制作的坐便圈,充分保证冬天如厕的舒适性;同时利用菌、酶、生物集成技术,达到粪便高效降解、消除恶臭等多重功效,减少的粪便量达到80%以上,产出的有机肥可直接作为基肥。同时,通过注塑工艺减少加工成本,使整套系统的销售价格低于250美元。

Designed for elderly households in rural China, this waterless toilet offers both squatting and sitting modes and features a feces and urine collection and resource utilization system. Grounded in the current conditions of rural China, it provides a user-friendly transitional toilet product with distinctive Chinese characteristics. The design preserves traditional squatting habits while meeting the growing demand for sitting toilets among the elderly. The product undergoes rigorous ergonomic testing and evaluation, using EPP (Expanded Polypropylene) foam for the toilet seat to ensure comfort during winter. Additionally, the system employs microbial, enzymatic, and biological integration technologies to efficiently degrade feces, eliminate odors, and reduce fecal volume by over 80%. The residual material after fermentation can be used as fertilizer, and reduces the cost of later fecal removal. By using injection molding technology helps lower production costs, making the entire system priced below $250.

SlaapLekker 健康监测仪计
SlaapLekker Health Monitor

杭州博博科技有限公司 | 中国设计智造大奖铜奖作品 | 2019
Hangzhou Bobo Technology Co., Ltd. | DIA BRONZE | 2019

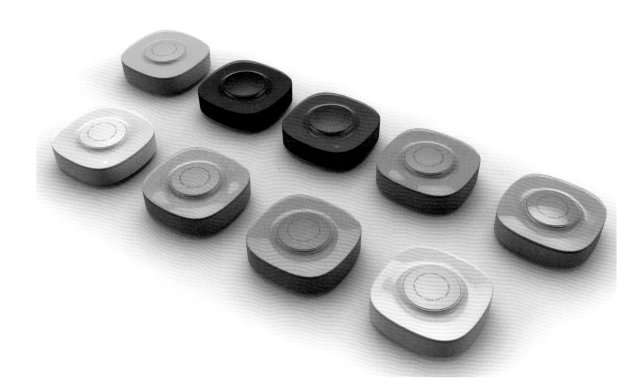

随着老龄化社会的到来，心脑血管疾病将不仅仅是一个健康问题，更是沉重的经济负担。世界卫生组织（WHO）指出心脑血管疾病所需的高昂治疗费用可通过早期发现和有效风险管理来规避。SlaapLekker是一款基于人工智能技术的无感知健康监测仪，用于早期发现心脑血管疾病的致病风险以及病情恶化，在为终端用户提供舒适的用户体验同时也兼顾了低廉的费用。它被认为是一种用于减少老龄化社会负担的有效方式，可以有效地改善患者的生存质量。

With the transition to aging populations, people all over the world will suffer more death and disability from cardiovascular disease. In the other hand, cardiovascular disease is no longer just a health issue, but a major economic burden. The World Health Organization indicates that cardiovascular disease is likely to require costly medical treatments that might have been avoided with early detection and effective management of risk factors. By integrating innovative artificial intelligence technologies. SlaapLekker is a non-obtrusive health monitor for the early detection of the cardiovascular disease risks or of its deterioration, such as higher continuous resting heart rate, atrial fibrillation and sleep apnea-hypopnea syndromes. As a home healthcare device, SlaapLekker aims to ensure a comforting user experience and the explainability of system intelligence to users, but with dramatic cost savings and improved cardiac care. The application of SlaapLekker is considered to an effective way to reduce the burden of cardiovascular disease for aging populations and improve the quality of life for patients.

益滴水
NiceDrip

郭倩，苏梓裔，王一淳，陆静雯 | 同济大学设计创意学院 | 2022
Qian Guo, Ziyi Su, Yichun Wang, Jingwen Lu | College of Design and Innovation, Tongji University | 2022

"益滴水"是一个未来医学概念，聚焦于有益健康的微生物群落对人体的影响。项目基于社区的大数据分析，通过人工智能赋能的菌群分析、健康模拟、干预模拟和自动配制，设计出家庭级的制菌机，可以根据个性需求生产不同菌种配比的可服用"水球"，对人体菌群进行积极干预，以增进身体健康和家庭福祉。项目同时认为，健康未来将从宏观层面扩展到微观层面，社区成为构建和共享微生物区系数据库的最小单位。作为一种家用产品，该系统的终端将使未来的精细微生物健康管理能够惠及数百万家庭。

NiceDrip is a futuristic medical concept that focuses on the impact of beneficial microbiomes on human health. The project integrates community-based big data analysis with AI-powered microbiome analysis, health simulation, intervention simulation, and automatic formulation to design a home-level bacterial production device. This device can produce consumable "water balls" with different microbial compositions tailored to individual needs, actively intervening in the human microbiome to promote health and family well-being. The project envisions that future health management will expand from the macro level to the micro level, with communities serving as the smallest unit for building and sharing microbiome databases. As a home product, the system aims to enable precise microbial health management for millions of households.

沟通界面如何展现出一种相互理解的创新文化？

卡路里能够测量出来，健康状况可以检查出来。但是，还有一个生活质量的关键指标缺少客观评价：信任。信任更多依赖于社会模式、直觉、信心和德性修养，它和文化一样是无形的。诚然，现有的国家法律法规、国际条约和全球组织机构或许可以为人们的相处方式提供框架规约，并由此产生一定程度的可预测性，但遵守规则的前提是人们对其心存信念，而这种信念则要由日常养成的信任来浇灌。设计是促成这种信任感的主要因素之一，它为人们提供灵感、钦慕感、自豪感、舒适感、好奇心、安全感、认知度及其他共同心态模式。长久以来，设计影响着我们与周围环境之间以及彼此之间进行交流的方式。好的设计是跨越各种信任危机的桥梁——不管这个危机的来源是阶层、种族、性别、文化、信仰、区域还是社会。设计是世界和平的润滑剂。

How can a communication interface represent a culture of innovation to mutual understanding?

Calories are measurable. Health can be checked. But there is another key quality of life that escapes such objectivation: trust. Relying much more on social patterns, intuition, confidence and the cultivation of virtues, trust is as intangible as culture can be. Granted, we may have national laws and regulations, international treaties and global institutions, that provide a framework for our behavior to each other, and provide a degree of predictability. But rules assume faith in them, and this faith requires daily cultivation of trust. Design is one of the primary enablers of such trust, offering people inspiration, admiration, pride, comfort, curiosity, a sense of security, recognizability, and other modes of common ground. Throughout our days, design frames our communication with our surroundings and with each other. Good design is a bridge that links various trust crises—regardless of whether the source of the crisis is class, race, gender, culture, belief, region, or society. Good design is a peacemaker.

城事设计节 | 美好新华
Urban Design Festival | Beautiful Life in Xinhua

AssBook 设计食堂 | 大鱼社区营造发展中心 | 2018
AssBook | Big Fish Community Design Center | 2018

近年来,关于"城市更新"的探讨非常热烈,但一般工作重点大多停留在针对物质空间的设计改造层面,真正能整合经济价值、经营模式、合作体系的成功案例并不多见。有鉴于此,"城事设计节"由 AssBook 设计食堂发起,试图通过"新组织 + 新设计 + 新媒体"的方式重新定义城市更新,让更多人参与建设并体验宜居城市生活,最终实现政府、设计团体、企业品牌以及在地居民的多赢局面。"城事设计节 | 美好新华"落地在上海新华路街道,选择老社区的典型设计专题,通过居民参与式设计工作坊、社区营造主题论坛等专题实践,探索上海老公房小区自治及管理机制,以真正激发一个城市社区基于人与人之间社会关系的积极互动而自然产生的活力。

In recent years, discussions on "urban renewal" have been very active, but the focus of most work has generally been on the design and renovation of physical spaces. Successful cases that truly integrate economic value, business models, and cooperative systems are rare. In light of this, the "Urban Design Festival" was initiated by AssBook, aiming to redefine urban renewal through "new organization + new design + new media". The goal is to involve more people in building and experiencing livable urban life, ultimately achieving a win-win situation for governments, design teams, corporate brands, and local residents. "Urban Design Festival | Beautiful Life in Xinhua" took place in Shanghai's Xinhua Road subdistrict, focusing on typical design topics of old communities. Through resident-participatory design workshops, community-building theme forums, and other practical activities, the project explored self-governance and management mechanisms of Shanghai's old residential neighborhoods, aiming to stimulate the vitality of urban communities through positive interactions based on social relationships.

云集城市
Swarm City

祝晓峰 | 山水秀建筑事务所 | 2022
Xiaofeng Zhu | Scenic Architecture Office | 2022

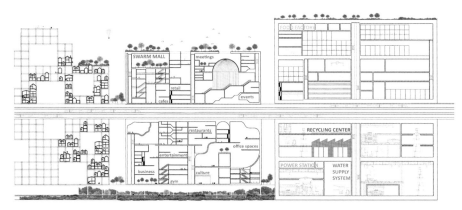

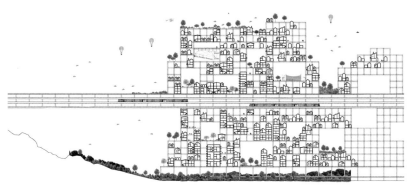

在高密度城市中考虑系统性的生态发展，其根本目标是推动人与自然的和谐共处。伴随着城镇快速扩张，生态空间不断受到挤压，如何应对城市空间需求和自然资源维护的矛盾，成为城市规划中最重要的挑战之一。"云集城市"是一个高密度城市原型，由地表自然、基础设施、住宅聚落和云集商业等部分组成，涉及城市群体生活与生态、农业、工业、信息、文化、卫生、教育等多个领域关系的革新。设计通过取消城市地面机动车及道路，把地表还给自然，把健康还给人和地球。

In high-density cities, systemic ecological development is fundamentally aimed at promoting harmonious coexistence between humans and nature. With the rapid expansion of urban and rural areas, ecological spaces are increasingly squeezed. Addressing the conflict between urban space demands and natural resource preservation has become one of the most significant challenges in urban planning. "Swarm City" is a prototype for high-density urban environments, composed of surface nature, infrastructure, residential clusters, and commercial areas. It involves innovations in relationships across various domains such as urban life, ecology, agriculture, industry, information, culture, health, and education. The design proposes eliminating surface motor vehicles and roads in urban areas, thereby returning the surface to nature and restoring health to both people and the planet.

"绿线"教学实践系统
"Green-Line" Artificial Ecological Symbiosis System

郭泠 | 上海市同济黄浦设计创意中学 | 2020
Ling Guo | Shanghai Tongji-Huangpu School of Design and Innovation | 2020

同济黄浦设计创意中学的"绿线"是一个用于教学与实践的人工生态共生模拟系统。它从屋顶收集雨水用于"城市农业"灌溉,通过开源硬件和程序控制的营养、湿度等微环境干预,使得室内的鱼菜系统和室外环境紧密相连。"绿线"不仅是一个景观设计,也是一个集教学、实践、游戏、创作和展示功能于一体的开源软硬件系统,更是一个巨大的互动的和不断增长的教具。本系统为"以问题为导向"的教学方法提供了支持,将真实世界问题嵌入学生的学校生活中,使教学隐形。通过持续实践,学生最终会理解人工系统的脆弱性与维护的重要性,这正是可持续发展的核心含义。

The "Green-Line" at Shanghai Tongji-Huangpu School of Design and Innovation is an artificial ecological symbiosis system used for teaching and practice. It collects rainwater from the roof for irrigation in "urban agriculture" and uses open-source hardware and program-controlled interventions for nutrients, humidity, and other micro-environmental factors, creating a close connection between the indoor fish-vegetable system and the outdoor environment. "Green-Line" is not only a landscape design but also an open-source hardware and software system that integrates teaching, practice, games, creation, and exhibition functions. It serves as a large interactive and ever-evolving teaching tool. The system supports a problem-based learning approach, embedding real-world issues into students' school lives and making teaching more invisible. Through continuous practice, students will ultimately understand the fragility and maintenance importance of artificial systems, which is the core meaning of sustainable development.

图腾椅
WDCC Tattoo Chair Series B No.3

设计师：约里奥·库卡波罗 | 1997
平面设计：杜钦 | 2022
品牌：阿旺特
Designer: Yrjö Kukkapuro | 1997
Graphic Designer: Qin Du | 2022
Manufacturer: AVARTE

西方的现代主义，无论是建筑还是其设计，始终保持着对图案式表达方式的热衷。通过平面和客观实体的设计变化来表现其永恒性和普遍适应性。1997年，约里奥·库卡波罗应邀重新改造他设计的赫尔辛基实用艺术博物馆的胶合板座椅，图腾椅便应运而生。它的结构非常简单，但有着引人注目的视觉效果。与平面设计师的合作，赋予了图腾椅独特的个性，从而变得更有意义和收藏价值。平面设计师杜钦与阿旺特公司合作，为首届世界设计之都大会主题展设计了一个特殊版本的图腾椅，其鲜明的色块组合为这个经典产品带来了全新视觉体验。

Western modernism, whether in architecture or design, has always held a strong affinity for patterned expression. It showcases its timelessness and universal adaptability through variations in flat and objective forms. In 1997, Yrjö Kukkapuro was invited to redesign the plywood seating for the Helsinki Museum of Applied Arts, leading to the creation of the Tattoo chair. Its structure is not only simple but also visually striking. For the inaugural World Design Capital exhibition, Qin Du, in collaboration with AVARTE, created a special version of the Tattoo chair, giving the classic product a fresh visual experience with bold color block combinations.

脸之书 2022
Book of Faces 2022

于尔约·索塔马, 潘剑锋, 杨利华, 佩卡·托伊瓦宁, 马蒂·海迈莱伊宁 | 2022

Yrjö Sotamaa, Jianfeng Pan, Emily Yang, Pekka Toivanen, Matti Hämäläinen | 2022

信任是合作的基石，设计则是信任的桥梁。无论是个人、组织，还是地区、国家，信任的建立和维系都与我们生活的各个层面息息相关。"脸之书"项目诞生于2012年，上海举行赫尔辛基设计首都系列活动之时。这本书采访了致力于中芬合作的42位代表人士，用他们的头像和故事串起了一个基于信任、合作和跨国友谊的网络。"脸之书2022"是向十年前的项目致敬的作品，我们追踪了42个面孔中的5个，并展示了他们十年后的近况和彼此的合作。本作品呈现了他们对设计与信任之间关系的最新思考，以及他们在设计实践领域的最新成果。

Trust is the cornerstone of cooperation, and design serves as the bridge of trust. Trust is the foundation of collaboration, and design serves as the bridge of trust. Whether at the personal or organizational level, or at the regional or national level, the establishment and maintenance of trust are closely related to various dimensions of our lives. Once trust is threatened, the fruits of our efforts will be deprived of nourishment. The project of "Book of Faces" originated when the World Design Capital Helsinki 2012 was held in Shanghai. Relying on the Tongji-Aalto design factory and the Sino-Finnish Center as the collaborative platforms, a group of friends from design, music, art and other creative fields had the opportunity to work together. This team consisted of members from Finland and China, and they established a network of trust and cross-cultural friendships. "Book of Faces 2022" is a fresh continuation of this project. It followed the journeys of 5 individuals out of 42 faces, showcasing their current situations ten years later. This work presents their latest thinking on the relationship between design and trust and their achievements in the field of design practices.

社区参与式博物馆
Community Participation Museum

舒克小组 | 大鱼社区营造发展中心 | 2021
Shuke Group | Big Fish Community Design Center | 2021

社区参与式博物馆是位于上海虹桥机场新村核心区域的社区空间，是一个鼓励分享、交流互动的社区活动空间，由大鱼营造团队策划并设计。它面向各年龄层的社区成员，大家可以在这里组织和参与活动，策划和参观展览，探讨和参与社区议题。馆内除了不定期更新与社区文化相关的主题展览，还设有一大三小的四个主要活动空间，关注社区的新老朋友可以在此组织各种活动和聚会。在社会参与式博物馆的空间里，每个社区成员都能享受舒适的环境，体验轻松愉悦的心境，感受有温度的社区。

The Community Participation Museum is located in the core area of Shanghai's Hongqiao Airport New Village. It is a community space designed for sharing, communication, and interaction. The museum caters to community members of all ages, offering a space for organizing and participating in activities, planning and visiting exhibitions, and engaging with community issues. The museum features a large and three small event spaces, hosting regular exhibitions related to community culture. It provides a comfortable environment where both new and longtime community members can organize various events and gatherings. In this participatory museum, every community member can enjoy a pleasant atmosphere, experience relaxation, and feel the warmth of the community.

创意人 100
CREATORS 100

新天地 | 2022
XINTIANDI | 2022

本项目是由新天地打造的创意平台。本项目认为人是创意社群最为宝贵的资源，通过每年与百余位先锋创意人对话交流、协作共创，为创意人及其作品提供更广阔的展现平台，在推动创意经济的同时，助力中国创意行业的可持续发展。

CREATORS 100 is a creative platform established by XINTIANDI. The project views people as the most valuable resource in the creative community. By engaging in dialogue and collaboration with over a hundred pioneering creative workers each year, it provides a broader platform for showcasing their works. The initiative aims to advance the creative economy while supporting the sustainable development of China's creative industry.

社交图层
Social Layer

社交图层 DAO | 2022
Social Layer DAO | 2022

社交图层是一个互相发放社交徽章的应用程序，通过构建信任网络支持社区营造的基础设施，能够帮助用户之间建立连接、产生认同。本应用程序将价值评判的主体由大型组织机构拓展到个体和小团队。用户可以在该应用程序中接收、查看和铸造徽章。每个徽章的名称、颁发原因、铸造者和接收者等信息被记录在不可篡改的区块链上。用户可以通过了解他人接收到的徽章的颁发原因等，来获取与他人建立信任的有效信息，以激发更多的创新行动。相较于只有既有团体可以发放徽章的项目，该应用程序更关注在社区建构过程中逐渐涌现出的社会关系。

Social Layer is an App that enables the exchange of social badges, fostering connections and a sense of identity. It builds a trust network as the foundation for community development. expanding value assessment beyond large organizations to individuals and small teams. Users can receive, view, and create badges, with details recorded on an immutable blockchain. Understanding the reasons behind others' badges promotes trust and spurs innovation. Unlike projects limited to established groups, it focuses on emerging social relationships during community building.

1-100-1000-10000：四叶草堂自然体验
1-100-1000-10000: Clover Nature Experience

上海四叶草堂青少年自然体验服务中心 | 2016—2022
Shanghai Clover Nature School Teenager Nature Experience Service Center | 2016-2022

"四叶草堂自然体验"项目实现了从 1 到 10 000 的飞跃式发展，涵盖了四个阶段。2016 年，项目在上海杨浦区创智天地园区内建成了首个位于开放街区的社区花园——创智农园，占地 2 200 平方米，标志着城市更新从空间资本化生产向社区营造的转变。2020 年，东明实验通过社区花园和街区规划，推动了社区微更新，促进公众参与的常态化。2022 年，南宁老友花园项目通过居民参与，在 155 个老旧小区内建成了 1 000 个小型社区花园。随后，团"园"行动通过全国性的公众参与，支持 10 000 名居民共建社区花园，推动了中国社区花园的发展，编制了全国社区花园白皮书和地图。

1：创智农园
上海首个位于开放街区的社区花园
2016

100：东明实验
从社区花园到街镇系统性社区规划
2020

1 000：南宁老友花园社区参与行动
市域层面专项社区空间更新
2022

10 000：团"园"行动
广域的公众参与自主行动网络
2022

The "Clover Nature Experience" project has achieved a remarkable leap from 1 to 10,000, spanning four distinct stages. In 2016, the project built the first community garden located in an open block within the KIC area in Yangpu District, Shanghai - KIC Agricultural Park, covering an area of 2,200 square meters, marking the transformation of urban renewal from spatial capitalization production to community building. In 2020, Blooming Dongming promoted micro regeneration and normalized public participation through community garden and block planning. In 2022, the Nanning Old Friends Garden project built 1,000 small community gardens in 155 old residential areas through resident participation. Subsequently, the "Group Garden" campaign supported 10,000 residents to jointly build community gardens through nationwide public participation, promoting the development of community gardens in China and compiling a national white paper and map of community gardens.

1: KIC Agricultural Park
The first community garden located in an open block in Shanghai
2016

100: Blooming Dongming
From community gardens to systematic community planning
2020

1,000: Nanning Old Friends Garden Community-based Public Participation
Special community space updates at the municipal level
2022

10,000: "Garden" Action in Group
Nationwide public participation and autonomous action
2022

为你做设计
User Generated Design

小红书 | 2022
Xiaohongshu | 2022

需求的普适性让普通人之间相互提供与汲取灵感,自然地做出了让生活更好的设计,而这些"好设计"每时每刻都在小红书里发生,真诚的社区让大家自然地看见彼此、相信彼此。基于这样的理念与观察,小红书首席产品官邓超从用户生成内容(UGC)的语境里向前提出了用户生成设计(UGD)的概念。

The universality of needs allows everyone to draw inspiration from each other, and naturally generate designs that make life better. And these "good designs" are happening in Xiaohongshu all the time—a genuine community where everyone organically sees each other and believes in each other. Based on such observation and deduction, Chao Deng, CPO of Xiaohongshu, proposes the concept of UGD (User Generated Design) within the context of UGC (User Generated Context).

信任与分享:刘毅的手机绘画与公共艺术
Trust and Share: Yi Liu Mobile Phone Painting and Public Art

刘毅 | 61CREATIVE | 2022
Yi Liu | 61CREATIVE | 2022

作为复星艺术中心天台艺术季第四季项目,艺术家刘毅带来了"从来没有,无聊的,时候",呈现了一系列延续自手机绘画系列的雕塑、装置及行为表演。被"屏幕蓝"色彩放大的手指雕塑,在外滩的天际线之间停留、漫步和跳舞。刘毅钟意于在这种多重宇宙之间的切换,数字作品与在地雕塑相互映衬,城市建筑与公共艺术交相辉映,这不仅是艺术家的创作,也是连同整个环境——云、天、光、建筑、江景和观众一起协同的创作。

As part of the fourth season of ART IN ELEVATION in Fosun Foundation, artist Yi Liu presents *Never, Boring, Time*, including a variety of sculptures, installations, performance art originated from the Mobile Phone Painting series. Fingers magnified by the shiny digital blue color, are wandering and dancing beneath the skylines of the Bund. The artist therefore enjoys the switchover among the multiverses, the echos of digital works and tangible sculptures, the conversation of city architecture and public art that enhance each other's beauty. This is not only a piece of art made by the artist, but a collaborated creation of the clouds, sky, light, terrace, river views and audience.

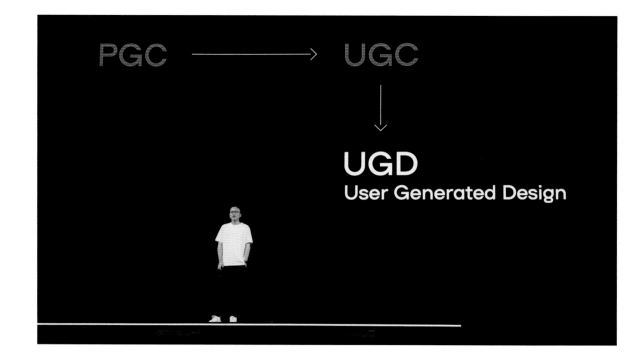

"燃冉"青年艺术家孵化计划
"RanRan" Young Artist Incubation Program

新天地,UCCA | 2022—2025
XINTIANDI, UCCA | 2022-2025

"燃冉"青年艺术家孵化计划是由瑞安新天地旗下商业品牌XINTIANDI新天地与UCCA共同发起的三年战略合作,旨在通过青年艺术奖项评选、艺术家驻留项目、艺术季、跨界合作四大创意板块为中国青年艺术家群体提供广阔的创意展示平台、开放的实验空间与自由的创作环境,同时为其提供跨界探索的多元可能性,并期待激发公众的文化好奇心与探索欲,构建可持续的艺术生态与文化场域,吸引和培养更广泛的文化艺术群体,推动中国艺术现场的发展。

The "RanRan" Young Artist Incubation Program is a three-year strategic cooperation established by XINTIANDI, the lifestyle brand of Shui On Xintiandi, and UCCA. It aims to leverage the resources and advantages of both parties and to provide a welcoming platform for creative endeavors, and an open and free environment for young Chinese artists to present their work and engage in interdisciplinary experimentation through four phases: Young Artist Prize, Artists-in-Residence Program, Art Seasons, and Crossover Collaborations. The program adopts a collaborative model that will work with young artists to explore wider possibilities of art, as well as further enhance the public's desire for and curiosity about culture, by building a vibrant public art and culture platform. By attracting the participation of cultural and artistic groups and helping cultivate them, it aims to promote the development of Chinese art in the future.

价值工厂
Value Factory

奥雷·伯曼 | UABB/CMSK | 1986—2013
Ole Bouman | UABB/CMSK | 1986-2013

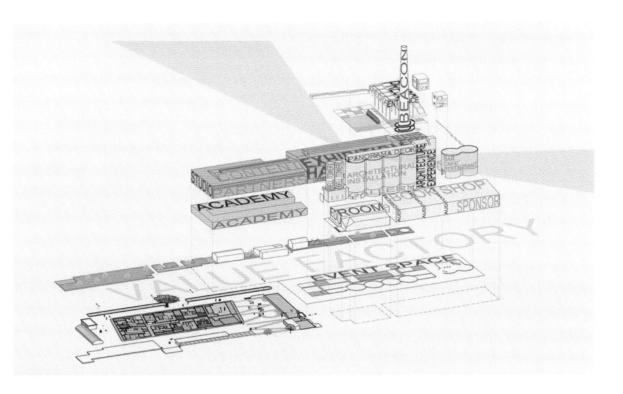

蛇口价值工厂是深圳旧工业遗址活化的代表项目,过去这里承载了20世纪80年代改革开放的记忆。价值工厂被选为2013年深圳双年展的举办地。从一开始,人们就说价值工厂的视野超出了双年展为期三个月的活动。它有望成为一个激发设计师、客户和来访市民大胆想法的长久的地方。创新的热土也是学习和实干的热土。2013年12月,当年的老玻璃厂重新打开,准备进行进一步的实验、使用和规划。如今,十年过去了,这个颇具创意的工业园区已成为老工业基地活化利用的标杆。

The Value Factory in Shekou is a representative project for revitalizing old industrial sites in Shenzhen, which once carried memories of the 1980s reform and opening-up era. The Value Factory was selected as the venue for the 2013 Shenzhen Biennale. From the very beginning, people said that the vision of the Value Factory extended beyond the three-month event of the biennale. It has the potential to become a long-lasting place that inspires designers, clients, and visiting citizens with bold ideas. This fertile ground for innovation is also a place for learning and practical work. In December 2013, this former glass factory was reopened, ready for further experimentation, use, and planning. Today, after a decade has passed, this creatively designed industrial park has become a benchmark for the revitalization of old industrial bases.

主题4: 环境
Theme 4: Environment

清洁行动如何展现一种创新文化以实现与自然的新平衡？

How can a cleaning act represent a culture of innovation to a new balance with nature?

自人类诞生以来，就需要通过食物、健康和信任来满足生存需求。对此我们时时有所感悟。相比之下，我们似乎经常忘记，惟有当自然生命世界也很健康，动植物得到滋养，地球生态系统得到妥善治理时，方能满足这些需求。如今，人地危机倒逼我们需要改变人类生活方式，以进入一个关爱自然、尊重自然、和谐共生的新时代。至少，设计正是一个思考替代运作模式、创意想法涌动的领域。从萃取到再生，从石油到土壤，从废物处置到循环利用，从单种栽培到生物多样性，从利润最大化到追求平衡。传统的"人本设计"思维，正在环境这个维度上呈现出新的内涵。

Food, health, and trust are requirements of life since the dawn of humankind. We never stopped realizing that. In contrast, periodically we seem to forget that these requirements can only be met if our natural lifeworld is healthy too, flora and fauna can be fed too, and if our planet can rely on our ecological stewardship. Currently, we are at the end of such an era of oblivion, and hopefully on the brink of a new era of care and respect for nature. Design, at least, is a field that brims with ideas for alternative modes of operation. From extraction to regeneration. From oil to soil. From disposal to recycling. From monocultures to biodiversity. From maximizing profit to cultivating balance. Good design brings measure to our life, and helps us appreciate it.

SHIELD：海浪助力式近岸垃圾收集设备
SHIELD: A Wave-assisted Nearshore Garbage Collection Equipment

严泽腾 | 清华大学美术学院 | 2021
Zeteng Yan | Academy of Arts & Design, Tsinghua University | 2021

据 2010 年统计数据，每年有多达 800 万吨垃圾被扔进海里。全球一半以上人口生活在沿海附近，近岸活动产生了大量垃圾。海岸线作为大陆与海洋的交界线，是阻止垃圾入海和外来垃圾上岸的一道重要防线。SHIELD 旨在对近岸垃圾起到收集和防御的作用，并以海浪作为动力实现垃圾收集。

According to 2010 statistics, up to 8 million tons of garbage are dumped into the sea each year. More than half of the global population lives near coastal areas, generating a significant amount of waste from nearshore activities. The coastline, as the boundary between land and sea, serves as an important defense against waste entering the sea and external waste coming ashore. SHIELD is designed to collect and prevent nearshore waste using ocean waves as the power source for garbage collection.

超级皮儿
PEELSPHERE

宋悠洋 | 朴飞生物 | 2021
Youyang Song | PEELSPHERE | 2021

使用环保材料造就我们周围的世界，正在成为设计的领先理念。例如时尚产业在探索皮革替代材料时，人们就看到了这一点。超级皮儿是一种百分百可循环利用的生物可降解材料，专注于水果废料与海藻的再利用潜能。本项目通过先进的材料工程，结合美学和设计原则，通过回收，重新设计，重新使用，将水果废料与海藻转化为美学与高性能并重的环保可降解生物材料，为皮革和合成皮革提供了理想替代品，实现了闭环的循环设计。超级皮儿取自自然，回归于自然。同时，本项目亦致力于发展成为材料开发的平台，推动更多低碳、结合设计的材料产品，具有巨大的潜力。

Using environmentally friendly materials to shape our surroundings is becoming a leading design concept, especially in the fashion industry where finding alternatives to leather is a key focus. PEELSPHERE is a 100% recyclable and biodegradable material that highlights the potential for reusing fruit waste and seaweed. This project employs advanced material engineering, integrating aesthetics and design principles to recycle, redesign, and repurpose fruit waste and seaweed into an eco-friendly, high-performance biodegradable material. It offers an ideal alternative to leather and synthetic leather, achieving a closed-loop design. PEELSPHERE originates from nature and returns to nature. Additionally, this project aims to evolve into a platform for material innovation, promoting more low-carbon, design-integrated material products, and holds significant potential.

循环生态屋
Circular & Ecology Pavilion

朱慧，冯新泉，罗瑶 | REDO 重塑设计 | 2020
Hui Zhu, Xinquan Feng, Yao Luo | REDO Design | 2020

本作品是循环园区改造项目的一部分，也是 REDO 在区域循环探索过程中的重要里程碑；设计主体框架使用 65% 稻谷壳再生仿木材料，屋顶及墙体隔断使用 40% 回收 PC 塑料制成的阳光板。本设计通过太阳能能源驱动厨余垃圾处理，将处理后的营养物质用蔬菜种植，种植出蔬菜食用后的残余物再次用于堆肥这一形式，试验并形成了厨余垃圾的"微观循环"。

This work is part of a larger circular park renovation project and represents a significant milestone in REDO's exploration of regional circularity. The design framework uses 65% recycled rice husk-based faux wood materials, while the roof and wall partitions are made from 40% recycled PC plastic sunboards. The design integrates solar energy to drive kitchen waste processing, which then uses the processed nutrients for vegetable cultivation. The residuals from the harvested vegetables are used to create compost, establishing a "micro-cycle" for kitchen waste.

HANQING DING 22AW 胶囊系列
HANQING DING 22AW Capsule Collection

丁罕青 | 2022
Hanqing Ding | 2022

本作品是2022年度RISE UP可持续时尚设计挑战赛的复赛作品。RISE UP可持续时尚设计挑战赛由R.I.S.E.可持续时尚创新平台于2021年首次发起，在面对气候变化和全球循环经济转型中应运而生，旨在挖掘并启发中国年轻一代具有前瞻视野、兼具商业价值和环境价值的可持续时尚顶尖设计人才。作品灵感来自服装开发期没有被最终使用上的废片，设计师将不完整的它们打乱重组，利用品牌特别的珠织工艺进行连接，创造了既随机偶然，又精心排布的新面貌。从这套服装上，我们能看到品牌这季所使用的颜色、图案、针织结构等。

This collection is a semifinal entry for the 2022 RISE UP Sustainable Fashion Design Challenge. The RISE UP challenge launched by the R.I.S.E. Sustainable Fashion Innovation Platform in 2021 was created in response to climate change and the global shift towards a circular economy. It aims to discover and inspire the next generation of top sustainable fashion designers in China who possess both forward-thinking perspectives and commercial and environmental value. The inspiration for this work comes from discarded fabric pieces that were made during the garment development phase. The designer rearranged these incomplete pieces in a random yet deliberate manner using the brand's special bead-weaving technique to connect them. This creates a new look that is both accidental and meticulously arranged. The collection showcases the colors, patterns and knit structures used by the brand this season.

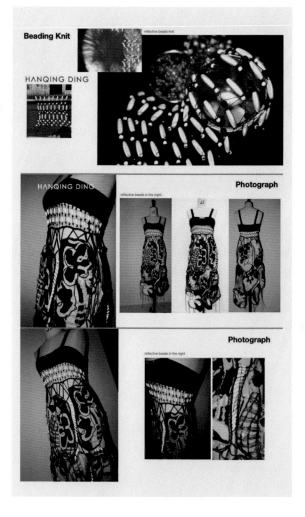

共生
ASSEMBLAGE

来雨晴，徐上茜 | 2021
Yuqing Lai, Shangqian Xu | 2021

本作品是 2021 年度 RISE UP 可持续时尚设计挑战赛优胜设计师品牌 MTG 的决赛作品。MTG 从共生（"ASSEMBLAGE"）的观念出发，两位设计师以花圃中的植物花卉为灵感，并结合了人类运动的元素：水上瑜伽。从一片园子出发，MTG 始终从一个地方或事物的角度设想未来。在 MTG 的花园中，这里所有生物的 DNA 都处在一种被"共享"的状态。在展出服装的制作中，MTG 选用可生物降解和堆肥，且含 40% 回收材料的 Naia ™ Renew 醋酸纤维素纤维制成的面料，并采用切割技术，将镂空后的花朵重新立体刺绣于衣服上，达到了零浪费的效果。

This work is a finalist entry from the 2021 RISE UP Sustainable Fashion Design Challenge, created by the winning designer brand MTG. Starting from the concept of "ASSEMBLAGE", MTG drew inspiration from the plants and flowers in the designers' garden, incorporating elements related to human movement, such as water yoga. MTG envisions the future from the perspective of a particular place or thing. In MTG's garden, the DNA of all living organisms exists in a state of "sharing". For the garment production showcased, MTG selected fabrics made from Naia ™ Renew acetate fibers, which are biodegradable, compostable, and contain 40% recycled materials. The design employs cutting techniques and features three-dimensional embroidery of perforated flowers on the garments, achieving a zero-waste effect.

宅
Being Home

侯正光，邱思敏，陈梓绵，李宣吉 | 晒上海 | 2020
Zhengguang Hou, Simin Qiu, Xuanji Li, Zimian Chen | Shine Shanghai | 2020

"晒上海"作为一项公益设计行为，自2009年起，以创意推动新材料和新工艺在设计中的应用，为需求做前瞻性研究。2020年"晒上海"的主题是"宅"。宅，是个动词，意为居住，窝着。宅，也是个形容词，用于描述居家的状态。近些年，"宅"更是作为一种亚文化而被广泛讨论。2020年，新冠病毒肆虐全球，人们被迫长时间宅在家中。当"宅"替代"在外奔忙"成为2020年初世界人民的集体经验之时，很多设计师都在进一步思考如何积极地"宅"，如何更好地构建"宅"，如何重新定义"宅"。以下四个作品正是从本次活动中产生，由极致盛放协同芬欧汇川（UPM）为本次活动提供了Formi材料及相应3D打印设备。

Formi材料是乳酸（PLA）和纤维素组成的复合塑料，是绿色环保的可持续性材料。同济大学设计创意学院则提供工坊的制作空间及设备，以及小批量注塑加工设备。

Since its inception in 2009, "Shine Shanghai" has been a public welfare design initiative that uses creativity to drive the application of new materials and techniques in design. The project conducts forward-looking research to meet demand. The theme of "Shine Shanghai" 2020 was "Being Home". In 2020, when the Covid-19 ravaged the world, people were forced to stay at home for long periods of time. As "Being Home" replaces "working outside" as the collective experience of the world's people in the early 2020, many designers were thinking about how to positively "Being Home", how to better "Being Home" and how to redifine "Being Home". The following four works were born in this event. Xuberance, collaborated with UPM, provided Formi materials and 3D printing equipments for this event.

Formi is a composite plastic made from polylactic acid (PLA) and cellulose fibers, which is an eco-friendly and sustainable material. College of Design and Innovation of Tongji University provided workshop space and equipment for small-batch injection molding.

"遇园" 桌凳
Yu Garden Table and Stool

侯正光 | 晒上海 | 2020
Zhengguang Hou | Shine Shanghai | 2020

"遇园"的设计灵感来自《芥子园》画谱里面的石谱。设计师取用了水墨石头的轮廓作为螺旋堆叠的基本型，产生了一个似是而非的巨石。本作品亦配有水纹玻璃，使其更像山顶的天池。很早以前，设计师就希望将阳台变成一块土地，不需奇花异草，只要杂草顽石。他将原本户外的巨石形态用3D打印的方式做成石凳石桌放在阳台，虽然其中没有泥土，却也别有一番景象，因为其重量远小于石头且方便移动。尽管我们不能拥有园林，但宅在家中亦可借此寄情书卷，神游山水，心中有园，故设计师为其起名"遇园"。

The design inspiration for "Yu Garden" comes from the stone patterns in the *Jieziyuan* painting manual. The designer takes the contour of ink-wash stones as the basic form for a spiral stack, creating an illusion of a massive rock. The design is equipped with water-patterned glass, which resembles a mountaintop celestial pool. The designer has long wished to transform the balcony into a piece of land, where flower might be absent but stones must be necessary. He used 3D printing to make stone-like stool and table for the balcony. They are much lighter than real stones and easy to move. Though it is hard to own a garden, staying at home allows us to immerse in literary scrolls and mentally wander through landscapes. Thus, the designer named this set "Yu Garden".

花塔：连接
CONNECT Vase

邱思涛，邱思敏 | 晒上海 | 2020
Sitao Qiu, Simin Qiu | Shine Shanghai | 2020

本作品创作于2020年全球疫情大爆发的背景下。一系列防控措施有效地遏制了病毒的传播，但人们被孤立在有限的空间里，渴望连接的重建显得如此珍贵。每一个花器都是独立的个体，它们层层相连，构成了一座花塔，从孔洞中冒出一枝枝鲜花，象征着美好的生命和希望。

This work was created during the 2020 pandemic when people were cut off from society and had to stay home. These vases can be freely connected into various patterns, symbolizing the breaking down of social isolation. When displayed together, the CONNECT vases remind us of how we were separated by the pandemic, yet through small efforts, we could still connect with the outside world and each other. The vases highlight the power of connection and shared experiences, even in difficult times.

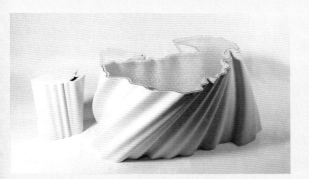

缠凳
Stool

李宣吉 | 晒上海 | 2020
Xuanji Li | Shine Shanghai | 2020

猫窝
The Home Life of a Pet Family

陈梓绵 | 晒上海 | 2020
Zimian Chen | Shine Shanghai | 2020

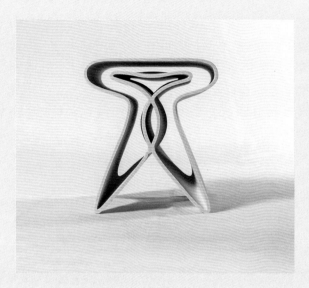

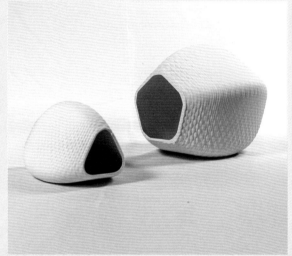

疫情期间,"宅"变成了一种生活的常态,在"宅"的状态下,当我们没有了更多的生活上的琐事,人最原始自然的行为与欲望变成的"宅"生活期间的主要状态,原始的欲望如两个线团相互纠缠,对于人宅在家里时的纠结与渴望,设计师采用了线的形式予以具象化。作品通过FDM方式的3D打印可回收木粉材料,通过线与线之间的相互挤压堆叠产生一种无限延伸的感觉,希望用作品唤醒人们在没有任何世俗的束缚下的思想解放。

对于一个养宠之家来说,可能需要花费额外的精力去购置宠物窝、玩具,本系列家具设计了适合宠物坐卧的空间,将人与宠物的使用需求结合,主人与宠物因为家具的共同使用又拉近了距离,增加了互动的机会。本设计将家具的功能性与装饰性结合——参数化算法生成的花纹复杂、不规律且独一无二,人们可以通过调整参数快速生产不同的定制产品,充分发挥了3D打印家具的快速一体化生产、针对不规则曲面的精准打印,以及与设计紧密结合的优势。通过机器人3D打印的可回收环保材料制造过程,做到了零废料零排放,实现了基于新材料新技术的环保制造。

During the pandemic, staying at home became a way of life. In this state of confinement, with fewer daily chores, the most primal natural behaviors and desires became central to life at home. In this work, the primitive desire is represented by two intertwined balls of yarn. The entanglement and longing experienced when people are confined at home are materialized through the form of yarn. The piece uses FDM 3D printing with recyclable wood powder material to create a sense of infinite extension through the pressing and stacking of lines. It aims to awaken people to the liberation of thought free from any worldly constraints.

For a household with pets, additional effort is often required to purchase pet beds and toys. This series of furniture takes pets' habits into account, integrating spaces suitable for pets to sit and lie down, and combines the needs of both people and pets. The shared use of the furniture enhances interaction and brings the owner and pet closer together. It merges functionality and aesthetics, with complex, irregular, and unique patterns generated by parametric algorithms. By adjusting parameters, different custom products can be quickly produced, leveraging the advantages of rapid, integrated production, precise printing on irregular surfaces, and close integration with design. The use of recyclable and eco-friendly materials in robotic 3D printing ensures zero waste and zero emissions, achieving environmentally friendly manufacturing based on new materials and technologies.

竹 C
Bamboo C

杨澍苗子，宋佳珈，杨洪君 | 北京服装学院 | 2022
Shumiaozi Yang, Jiajia Song, Hongjun Yang | Beijing Institute of Fashion Technology | 2022

本项目是以改良传统竹编为方向的创新型设计。围绕现代社会中竹编文化发展困难这一现象，希望通过简化竹编步骤、增加趣味性来使更多人接受竹文化，让竹产品重新融入人们的生活。传统手工竹编因其烦琐的编织过程和复杂的编织手法令大众望而却步。本设计重新定义竹材料的组合方式，采取了模块化设计原则，用简单的排列取代编织，由配套组件完成固定。由此，可以组成无限延伸的立体结构，给用户提供更多创意空间。

This project represents an innovative design approach aimed at improving traditional bamboo weaving. Addressing the challenges faced by bamboo weaving culture in modern society, the work aims to simplify the weaving process and increase its appeal, making bamboo culture more accessible and reintegrating bamboo products into daily life. Traditional handcraft bamboo weaving often deters the public due to its intricate processes and complex techniques. This design redefines the combination of bamboo materials by adopting a modular design principle, replacing weaving with simple arrangements, and using matching components for fixation. This allows for the creation of infinitely extendable three-dimensional structures, providing users with greater creative freedom.

祎设计再生砖
Yi Design Recycled Bricks

祎设计 | 2021—2022
Yi Design | 2021-2022

祎设计利用回收陶瓷废料研发出可用于陶瓷生产的泥料,包括适用于室外铺地的回收陶瓷再生透水砖,以及用于室内外的墙面装饰砖和手工砖等。

YiBrick - 陶瓷再生透水砖

回收率达90%以上,每平方米5cm厚的透水砖回收53kg以上的陶瓷废料,透水系数达7.74×10^{-2} cm/s。作为一种渗透性材料,这种多孔陶瓷砖适用于海绵城市路面,有助于缓解城市内涝、水资源短缺和热岛效应。

YiTile - 陶瓷再生瓷砖

回收率在70%以上,每平方米回收25kg以上的陶瓷废料。通过调配废瓷以及其他工业固废的颗粒度和添加量,经过二次节能烧制后可以广泛应用于外墙装饰、内墙装饰、地面装饰等领域。

YiBrick - 陶瓷再生手工砖

回收率达80%。可适用于室内和室外的墙面装饰。祎设计的独特美学追求,使其在众多的材料再生品牌中脱颖而出。

Yi Design has developed clay materials from recycled ceramic waste for use in ceramic production. This includes recycled ceramic permeable bricks suitable for outdoor paving, as well as decorative wall tiles and handcrafted bricks for both indoor and outdoor use.

YiBrick - Regenerated Ceramic Permeable Brick
Recycling rate over 90%. Each square meter of 5 cm thick permeable brick recycles over 53kg of ceramic waste, with a permeability coefficient of 7.74×10^{-2} cm/s. As a permeable material, this porous ceramic brick is suitable for sponge city pavements and helps alleviate urban flooding, water scarcity, and the urban heat island effect.

YiTile - Regenerated Ceramic Tile
Recycling rate over 70%. Each square meter recycles over 25kg of ceramic waste. By adjusting the particle size and amount of waste porcelain and other industrial solid wastes, and through a secondary energy-saving firing process, these tiles can be widely applied in exterior wall decoration, interior wall decoration, and flooring.

YiBrick - Regenerated Ceramic Handmade Brick
Recycling rate up to 80%. This handmade brick is suitable for both indoor and outdoor wall decoration. Yi Design's distinctive aesthetic approach makes it stand out among many technologically driven material recycling brands.

"山君"系列
"Gentleman in the Mountain" Series

洪丽娟,李明亮 | 无山 | 2022
Lijuan Hong, Mingliang Li | Seek Range | 2022

本作品描绘了人类与自然的连接关系与状态。人类从石器时代到工业社会,与自然的关系由亲近到疏离,再到由于自然环境危机的重新定义。本系列中的家具与家居用品由菌丝材料制作而成,摆放在家中,用于人类日常生活的一部分。在小柜子的设计中,设计师预留了可以结合植物种植的空槽结构,并融入柜子与花盆的功能——建立"长在家具上的植物"理念。这形成了一种关上门的亲近关系,建立起一种生活与自然息息相关,默默相伴的连接与情感。

This work explores the evolving relationship between humanity and nature, from the Stone Age to the industrial era, and towards a renewed connection driven by environmental crises. The furniture and home goods in this series are made from innovative mycelium materials. The design incorporates spaces for planting, blending cabinet and planter functions, and embodies the concept of "plants growing on furniture". This approach fosters a close relationship with nature, reflecting a lifestyle where natural elements are integrated into daily living.

设创学院的 1001 盏灯
1001 Lights in D&I

容晓薇，徐艺晨，陈鸣惊，熊天琦 | 同济大学设计创意学院 | 2022

Xiaowei Rong, Yichen Xu, Mingjing Chen, Tianqi Xiong | College of Design and Innovation, Tongji University | 2022

公共建筑有着较高的能耗水平，也具有特殊的社会角色和服务属性。本作品的设计小组从同济大学设计创意学院大楼出发，通过翔实的调研，探讨楼内照明系统设计和用户行为的可持续性，并设计了一套基于用户行为习惯的智能节能照明系统。本系统设计了新的照明开关交互方式，以及娱乐化的数字互动界面。设计师们相信：人让环境更智慧——基于对节能行为的引导，可以形成更加可持续的节能影响，同时可以对全社会起到示范引领作用，从而产生重要的经济、环境和社会效益。

This project addresses the high energy consumption and unique social roles of public buildings. The design team investigated the lighting system and user behavior in the College of Design and Innovation building at Tongji University. They developed an intelligent energy-saving lighting system based on user habits, featuring new lighting switch interactions and engaging digital interfaces. The system aims to promote energy-saving behaviors, making environments smarter and setting an example for broader societal sustainability, with significant economic, environmental, and social benefits.

以人为本的城市绿化
People-Centered Urban Green

魏佛兰，法比恩·普芬德，孙秀林，陈伟 | 上海大学中欧工程技术学院，同济大学设计创意学院，上海大学数据科学与都市研究中心 | 2020

Francesca Valsecchi, Fabien Pfaender, Xiulin Sun, Wei Chen | Shanghai University, Costech – Université de Technologie de Compiègne, College of Design and Innovation, Tongji University, Center for Data and Urban Sciences, Shanghai University | 2020

本项目利用数据科学，将城市绿色基础设施视为城市与市民之间的动态关系，而非静态规划结果。地图展示了不同地区和不同程度的城市绿化情况，直观呈现绿色发展在普通通勤和生活模式中的位置。绿色基础设施涵盖从游憩景观到生产性用地的各种功能空间，包括从完全自然到完全人工的要素。绿色基础设施通常被定义为环境中拥有固有性质、目的和规模的静态物体，但此定义未能描绘其对城市社会的影响。设计师们提出将绿色基础设施作为城市居民生活的重要组成部分，通过评估上海的5 000个社区，详细了解每个基础设施的性质及它们与居民的互动，从动态角度引导规划和设计方案。

This project utilizes data science to view urban green infrastructure as a dynamic relationship between the city and its residents, rather than a static planning result. The project maps various levels of urban greening, presenting the role of green development within daily commuting and living patterns. Green infrastructure encompasses a range of functional spaces from recreational landscapes to productive land, incorporating elements from fully natural to entirely artificial. Traditionally defined as static objects with inherent properties, purposes, and scales, green infrastructure has not been adequately described in terms of its impact on urban society. This project proposes considering green infrastructure as an integral part of urban life, assessing 5,000 communities in Shanghai to understand the nature of each infrastructure and its interaction with residents, guiding planning and design from a dynamic perspective.

基于可持续的游泳馆水循环系统研究
Research on Water Circulation Systems in Swimming Pools Based on the Sustainability Concept

李玉玺，田一清，朱昕怡 | 同济大学设计创意学院 | 2020
Yuxi Li, Yiqing Tian, Xinyi Zhu | College of Design and Innovation, Tongji University | 2020

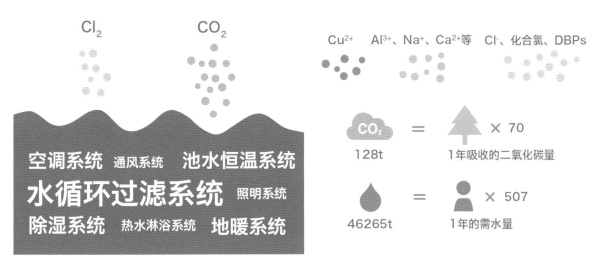

游泳作为一项全民运动深受大众的喜爱，尤其在烈日炎炎的夏季。在游泳馆的运行过程中，由于人体自然排出的污染物等原因，水需要不断进行循环消毒，以达到国家卫生标准，在这个过程中，游泳馆的耗水、耗电量十分可观。本研究基于可持续的理念及理论，将游泳馆的水循环系统进行系统分析解剖，将循环系统中水的去向及消毒剂的使用进行可视化的表达，并提出更"健康"的游泳馆使用方式。

Swimming is a popular exercise enjoyed by many, especially in hot summer months. However, swimming pools require constant water circulation and disinfection to meet national hygiene standards due to contaminants from swimmers. This process leads to significant water and energy consumption. This research, grounded in sustainability principles, conducts a systematic analysis of swimming pool water circulation systems. It visualizes the flow of water and the use of disinfectants within the system and proposes healthier usage practices for swimming pools.

竹林间
Bamboo Forest

章俊杰 | 素生设计机构，万华化学集团股份有限公司 | 2019
Junjie Zhang | SOZEN Design, Wanhua Chemical Group Co., Ltd. | 2019

本作品采用交叉曲线设计语言，采取模块化的参数设计理念，利用材料自身的刚性，分块组合形成一体化的框架，它既能体现结构美感，也能形成自然植物的向上生长的动态。经过单体不断的拼接延展后，宛如具有弹性的一片竹林，在山谷微风吹拂下，轻柔摇曳。人们在其下穿梭，能看到展品渐影渐现，拉长并丰富了视觉感受，让人在景中游，在景中觅，具有更强的互动性。作品材料是由万华公司提供的可回收塑料。

This work utilizes an intersecting curve design language and modular parametric design principles. It employs the inherent rigidity of materials to form an integrated frame through sectional assembly. The design not only highlights structural aesthetics but also creates a dynamic effect resembling the upward growth of natural plants. After continuous modular extension, the structure resembles a flexible bamboo forest that gently sways in the mountain breeze. Visitors moving through this space experience a visual transformation as the exhibits gradually emerge, enhancing the interactive and immersive experience. The materials used are recyclable plastics provided by Wanhua Chemical Group.

蜜蜂屋
Bee Home

巴肯与拜克，塔尼塔·克莱因，SPACE10 | 2020
Bakken & Bæck, Tanita Klein, SPACE10 | 2020

蜜蜂对花卉、树木、动物和人至关重要。我们日常食物中有三分之一有赖于这种昆虫的传粉效应。由于人类的影响，它们的家园和自然栖息地正在遭到破坏。蜜蜂正面临灭绝的危险，杀虫剂、化学品和单一栽培都对它们的生存构成了巨大威胁。这个简单的蜜蜂屋设计是开源的，让任何人都可以设计、定制和生产自己的蜜蜂屋。这是一种帮助地球茁壮成长的便捷方式，为蜜蜂提供了更多选择，让它们在我们自己的花园内得到安全的庇护。

Bees are vital for flowers, trees, animals and people. One third of what we eat depends on these types of insects and pollinators. Because of human impact, their homes and natural habitats are being destroyed. Bees are in danger of going extinct, with pesticides, chemicals and farming monoculture all posing great threats to their existence. This simple Bee Home design is open source, allowing anyone to design, customize and produce their own Bee Home. This is an accessible way to help the planet Earth thrive by providing bees with more options for safe shelter within our own gardens.

唤醒
Awaken

DESIGN SOCIAL | 新天地 | 2021
DESIGN SOCIAL | XINTIANDI | 2021

本作品旨在"唤醒"众人对濒危鸟类的保护和可持续时尚的关注。这是DESIGN SOCIAL by XINTIANDI联合GreenChallenge可持续时尚大赏的9位中国设计师（陈丹琪、李芬芬、李伟刚、严瑾、陆玺娅、单晓明、黄婉冰、梦会停、沈威廉）发布的首个数字艺术系列作品。通过数字科技的赋能，本作品意图唤起人们对环境问题的关注和对中国原创设计的版权保护意识，在设计创新领域谱写属于时尚设计与环境保护的数字文化故事。

This work aims to "Awaken" public awareness of endangered bird protection and sustainable fashion. In collaboration with nine Chinese designers (Danqi Chen, Fenfen Li, Weigang Li, Jin Yan, Xiya Lu, Xiaoming Shan, Wanbing Huang, Huiting Meng, and Weilian Shen) from the GreenChallenge Sustainable Fashion Awards and supported by DESIGN SOCIAL by XINTIANDI, the project introduces its first digital art series. Through the empowerment of digital technology, it seeks to raise awareness of environmental issues and advocate for copyright protection of original Chinese designs, creating a digital cultural narrative that merges fashion innovation with environmental conservation.

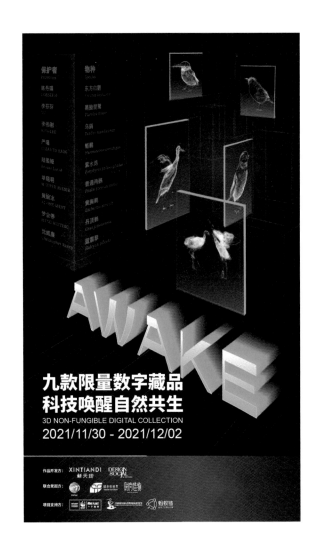

电能共生体
Power Mutualism

胡可儿 | 同济大学设计创意学院，清华大学深圳国际研究生院 | 2021

Kerr Hu | College of Design and Innovation, Tongji University; Tsinghua Shenzhen International Graduate School, Tsinghua University | 2021

长期以来，人类在从大自然中获取能源的同时，也对大自然造成巨大的伤害。随着电子产品向更轻巧、移动性更强、更可穿戴的趋势发展，我们生产电、使用电的方式需要被重新设计。电能共生体是一种可穿戴自驱动型蓝藻细菌发电材料，能够从人体汗液、阳光和空气中获取养分，通过蓝藻细菌光合和呼吸作用释放电能，为可穿戴小型智能设备提供绿色能源，使其摆脱外部电源充电、需要更换电池等限制，满足近未来可穿戴传感网络与物联网的分布式能源需求。

For a long time, while humanity has been extracting energy from nature, it has also inflicted significant damage on it. As electronic products evolve towards being lighter, more portable, and wearable, our methods of generating and using electricity need to be rethought. Power Mutualism is a wearable, self-powered material made from cyanobacteria that can harvest nutrients from human sweat, sunlight, and air. Through the photosynthesis and respiration of cyanobacteria, it generates electricity, providing green energy for small wearable smart devices. This innovation eliminates the need for external power sources, battery replacements, and supports the distributed energy demands of future wearable sensor networks and the Internet of Things (IoT).

主题5: 技术
Theme 5: Technology

人体的新延伸如何展现人类能动性的创新文化？

How can a communication interface represent a culture of innovation to mutual understanding?

人类在人与自然的平衡、人与人的平衡以及个人内心的平衡方面，无论其发挥的价值有多少，总是以干预和介入的方式凸显自己。某一刻看似正确和审慎的抉择，下一刻可能看起来有所欠缺，不尽人意，甚至无聊至极。文明的要旨是追求精益求精，而我们总是利用各种工具不断改进。但这种以人为本的方法已经为技术创新所替代，随着与人类能动性相混合，新技术日渐显现出一种自我监控能力。今天，技术并非仅仅是工具。它已成为人类智慧的伙伴。人工智能、算法应用、机器人和机器学习，以及基因工程等都是技术的具体方面，一如我们主导着技术，这些技术也对我们如何定义自身产生越发深刻的影响。技术本身改变着我们的自我意识和自我认知。

Whatever the value of finding balance with nature, with each other and with ourselves, humans manifest themselves by interventions. What at some point seems right and measured, the next moment may look insufficient, unsatisfactory, or even completely boring. Civilization was about the pursuit to improve, and we have always used our tools to make things better. But this human centered approach has given way to new technologies, increasingly showing a self-monitoring capacity in a hybrid mix with human agency. Technology today is not just about tools. It has become a partner of human ingenuity. Artificial intelligence, algorithmic applications, robotics and machine learning, and genetic engineering are all features of a technology that increasingly determines us as much as the other way round. This trend has an increasingly profound impact on how we define ourselves. Technology itself changes our sense of self and self-cognition.

英雄臂
Hero Arm

Open Bionics | 中国设计智造大奖金奖作品 | 2019
Open Bionics | DIA GOLD | 2019

随着文明和科技的进步，社会对残疾人士的关注度逐渐增加。尽管市面上有很多机械假肢，但高昂的价格、复杂的系统、异类的外观让很多人望而却步。本产品是一种多握把仿生手，以高性价比提供市场领先功能，适用于8岁及以上的人群。作为世界上首个3D打印的仿生臂，它不仅从技术上实现了假肢的各种基本功能，更通过人工智能和3D打印的结合，为人们提供了无穷的个性化选择，让拥有这个假肢的人因为自己与众不同的创造力和选择而感到骄傲。

With the development of civilization and technology, people with disabilities are getting more and more attention. Although there are many mechanical prosthetics on the market, their high price, complex systems, and exotic appearance are prohibitive for many people who really need it. The Hero Arm is a multi-grip bionic hand. It offers functionality and is now available through prosthetics clinics. The Hero Arm is the world's first 3D-printed bionic arm and the first to be available for children as young as 8 years old. It not only provides the essential functions of a prosthetic but also combines artificial intelligence with 3D printing to offer limitless personalization options. This allows users of the Hero Arm to take pride in their unique creativity and choices.

X-Craft：增强现实眼镜
X-Craft: Augmented Reality Glasses

杭州灵伴科技有限公司 | 中国设计智造大奖铜奖作品 | 2021
Hangzhou Lingban Technology Co. Ltd | DIA BRONZE | 2021

各种科技的应用拓展了我们的身体能力，如对人类听觉、触觉和视觉的增强。与此同时，我们的感官也发生了根本性的变革。在特定的专业领域，如矿井油田里的工人，最主要的工作痛点总结起来有两个：如何解放双手以及更安全。X-Craft是全球首款AI+AR+5G防爆现实增强智能眼镜，也是助力产业升级工业互联的智能化平台。它集成了混合现实显示、人工智能算法和5G传输，为工人赋予超强感知力，大大提高了工人的工作效率。可用于电力、石化、燃气、航空、轨交、运输、消防、基建、半导体等行业。

The application of various technologies has expanded our physical capabilities, enhancing our auditory, tactile, and visual senses. Concurrently, these advancements have brought about fundamental changes in our sensory experience. In specific professional fields, such as workers in mines and oil fields, two main challenges are identified: freeing up hands and ensuring safety. X-Craft is the world's first AI+AR+5G explosion-proof augmented reality smart glasses, serving as an intelligent platform for industrial upgrades and industrial connectivity. It integrates mixed reality displays, artificial intelligence algorithms, and 5G transmission, endowing workers with enhanced sensory perception and significantly improving their work efficiency. X-Craft is applicable in industries such as power, petrochemicals, gas, aviation, rail transit, transportation, firefighting, infrastructure, and semiconductors.

toio™ 专用主题（工作生物 Gesundroid）
Papercraft Creatures-Gesundroid

索尼互动娱乐有限公司 | 中国设计智造大奖金奖作品 | 2020
Sony Interactive Entertainment LLC | DIA GOLD | 2020

Gesundroid是机器人玩具"toio"的专属称号。通过将使用封闭的工作表创建的纸艺与两个可以移动的电动 toio Core Cubes相结合，并使用控制程序，用户可以构建可以剧烈移动并且似乎还活着的"纸艺生物"。

Gesundroid is an exclusive title of the robot toy "toio". By combining papercraft created using enclosed worksheets with two motorized toio Core Cubes that can move around, and using a control program, users can build "paper craft creatures" that can move about vigorously and seem to be alive.

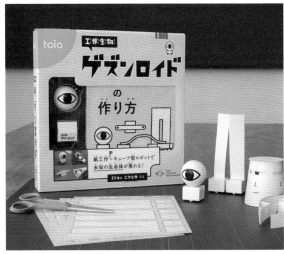

VINCI：智能海报设计小程序
VINCI: A Mobile App for Intelligent Poster Design

金卓宸，李静文，李照睿，石洋，曹楠 | 同济大学智能大数据可视化实验室 | 2021

Zhuochen Jin, Jingwen Li, Zhaorui Li, Yang Shi, Nan Cao | Intelligent Big Data Visualization Lab (iDVx Lab), Tongji University | 2021

VINCI运用前瞻科技，率先将最前沿的深度学习技术运用到平面设计领域，颠覆了传统的海报设计行为，将智能设计应用引入大众视野。在设计上，VINCI通过模拟专业设计师的思维模式和设计行为，遵循定制化的设计原则，采用一键式的生成方式，带给用户高效便捷的设计体验。作为科技与设计的结合，VINCI极大降低了设计领域高门槛的专业限制，使无专业背景的用户也可在20秒内制作一张精美海报，将"人人都能成为设计师"的理念变为现实。

VINCI utilizes cutting-edge technology to revolutionize poster design by integrating advanced deep learning techniques into the field of graphic design. This innovation disrupts traditional design practices and introduces intelligent design applications to the general public. VINCI emulates the thought processes and design behaviors of professional designers, adhering to customized design principles with a one-click generation feature, providing users with an efficient and convenient design experience. By combining technology with design, VINCI significantly lowers the professional barriers in the design field, allowing even users without a professional background to create stunning posters within 20 seconds, turning the concept of "everyone can be a designer" into reality.

人工智能赋能非遗传承：金山农民画
Art and Craftsmen Preservation Empowered by Artificial Intelligence: Jinshan Farmer's Painting

设计人工智能实验室 | 同济大学设计创意学院 | 2019
Design A.I. Lab | College of Design and Innovation, Tongji University | 2019

同济大学设计人工智能实验室研究如何通过人工智能技术进行非物质文化遗产的传承。最近三年，实验室建立了金山农民画的数据集、智能生成模型、人机交互模型，来探讨智能技术如何赋能传统农民画艺人对文化遗产的传承和发展，并提升用户的参与感，让非遗传承更具有时代特征。通过把创作过程、风格、结果转译成数字资产——从元素、主题、布局等维度对金山农民画进行数据化和原子化，为工艺美术的学术研究提供了新的维度。

The Design A.I. Lab investigates how AI technology can aid in the preservation of intangible cultural heritage. Over the past three years, the laboratory has developed a dataset, intelligent generation models, and human-computer interaction models for Jinshan Farmer's Painting. The research explores how smart technology can empower traditional folk artists to preserve and develop cultural heritage, enhancing user engagement and adding contemporary features to heritage preservation. By translating the creative process, style, and outcomes into digital assets — through datafication and atomization of elements, themes, and layouts — this work provides a new dimension for academic research in arts and crafts.

移动木结构加工机器人
Mobile Timber Construction Robot

袁烽 | 同济大学 | 2020
Philip F. Yuan | Tongji University | 2020

建筑机器人现场建造是智能建造领域研究的热点话题。本作品是移动建筑机器人的研究成果,可以实现机器人在建筑施工场地内进行木结构的搭建作业。

On-site construction by robots is a prominent research topic in the field of intelligent construction. This work represents the research outcomes of a mobile construction robot capable of performing timber structure assembly tasks on construction sites.

碳纤维复合结构板
Carbon-Fiber Composites Structural Panel

梁喆 | 上海大界机器人科技有限公司 | 2020
Zhe Liang | RoboticPlus.AI | 2020

本产品为承德阿那亚禅堂初次材料试验样品。禅堂是金山岭上院的重要组成部分，项目由阿那亚投资开发，整体方案由大舍建筑事务所设计，是大界智造联合和作结构建筑研究所完成的中国第一座，也是最大一座使用全碳纤维屋面的展亭。项目位于河北承德五道口，地处金山岭长城边的群山峻岭之中，交通极为不便。项目的亮点是使用碳纤维夹心屋面结构及其生产预制化，现场快速吊装。超轻的屋面材料重量，极薄的屋面结构层、防水层、装饰层，三层合一压缩在35mm的厚度之中。

This product is the first material sample of Zen Hall, Upper Cloister in Aranya, Golden Mountain. This beautiful resort next to the Great Wall was developed by Aranya, the Architect of Zen Hall is ATELIER DESHAUS, who collaborates with AND (structure engineer) and RoboticPlus.AI (robotic fabrication). It is the first and largest pavilion constructed with an all-carbon-fiber roof structure. Located in Wudaokou, Chengde, Hebei Province, in the middle of mountains and mountains, traffic is comparably inconvenient, in which we creatively use an extremely thin carbon-fiber composites product with a sandwich structure, integrating the structural layer, waterproof layer, and decorative layer within 35 mm thickness.

城市视景平台：数据驱动的基于实证分析的城市决策支持系统
CityScope: A Data-Driven & Evidence-Based Urban Decision-Support System

刘洋，王灿，葛霁雯，肯特·蓝森，马库斯·埃尔卡沙，卡森·斯姆，托马斯·桑切斯·伦格林，路易斯·阿尔贝托·阿隆索·帕斯托 | 同济－麻省理工上海城市科学实验室 | 肯特蓝森科技（上海）有限公司 | 2020

Yang Liu, Can Wang, Jiwen Ge, Kent Larson, Markus Elkatsha, Carson Smuts, Thomas Sanchez Lengeling, Luis Alberto Alonso Pastor | Tongji-MIT City Science Lab @ Shanghai | Kent Larson Technology (Shanghai) | 2020

深圳城市视景平台位于深圳市规划与自然资源局模拟实验中心，是以香蜜湖为研究区域的AI辅助城市设计决策支持实验室。平台覆盖城市设计、数据科学、人机交互、社会科学等领域，为城市设计决策提供动态的、可迭代的、基于实证的、多方参与的新范式。平台将设计思维引入城市决策支持系统，提升其实用性及易用性，整合从方案生成、分析模拟、可视化反馈、讨论、方案优化的城市设计决策全流程。平台实现了动态辅助城市设计科学决策的美好愿景，展现了具有示范意义的工作理念、方法和路径。

The Shenzhen CityScope platform, located at the Simulation Laboratory of the Shenzhen Planning and Natural Resources Bureau, serves as an AI-assisted urban design decision-support lab focusing on the Xiangmi Lake area. This platform spans urban design, data science, human-computer interaction, and social sciences, offering a dynamic, iterative, and evidence-based paradigm for urban design decision-making. By integrating design thinking into the urban decision-support system, it enhances both usability and functionality. The platform covers the entire urban design decision-making process — from proposal generation, analysis simulation, and visual feedback to discussion and optimization. It embodies the ideal of dynamically supporting scientific urban design decisions, showcasing a model approach, methodology, and pathway with significant demonstration value.

矩阵
The Matrix

郑康奕 | 同济大学设计创意学院 | 2022
Kangyi Zheng | College of Design and Innovation, Tongji University | 2022

自然孕育了生命，使其繁衍、生长，从而诞生了智慧。经历数万年的演化，世界展现了无尽的多样性，而人类正逐步迈入数字化时代，文字、声音、影像通过数字信息传递并记录。而机器又是否进一步产生智慧？本作品以造物主的视角，描绘赛博空间中万物与数字的关系。矩阵既为母体，作为信息的载体，记录存储着万物的形态和生长的规律，同时也代表着文明和希望。通过人工智能算法产生矩阵，展现生命具有多样性、随机性与可编辑性。作品为实时机械互动装置，由计算机随机生成点阵，映射在有机溶液中。过程中均由机器自主选择、判断位置和大小。最终矩阵图形呈现了数字艺术生成的完整过程。

Nature nurtures life, leading to wisdom through millennia of evolution, while humanity enters the digital era, recording and transmitting information. This work explores whether machines can generate wisdom. Viewed from a creator's perspective, it portrays the relationship between digital and physical realms in cyberspace. The matrix, as both a medium and symbol, stores the forms and growth patterns of all things, representing civilization and hope. Using AI algorithms, the matrix showcases life's diversity, randomness, and editability. This real-time mechanical installation features a computer-generated dot matrix projected into an organic solution, with the machine autonomously determining positions and sizes, revealing the digital art creation process.

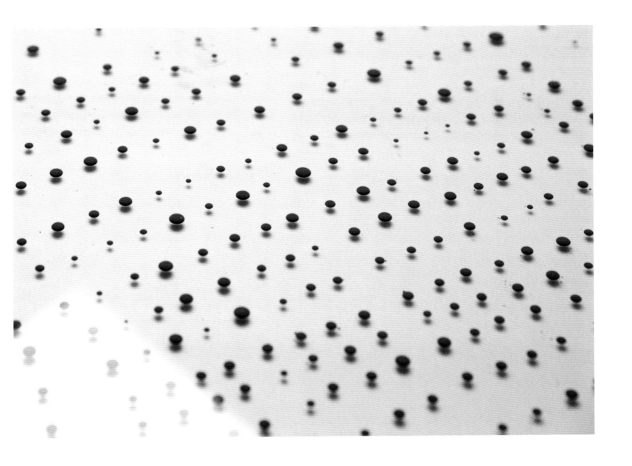

主题6: 繁荣
Theme 6: Prosperity

替代性的经济秩序如何展现创新文化对繁荣的新定义？

How can an alternative economical order represent a culture of innovation to a new definition of prosperity?

长久以来，我们通过经济增速、资本积累、薪资收入、出行距离、消费水平、出行方式、房屋面积、私有物品来衡量我们的进步。许多人认为这种从物质层面定义的财富等同于繁荣，并以此为动力，朝同一方向不断前进。让我们面对现实：设计是导致这种范式盛行的关键因素。

今天，繁荣的概念正在发生深刻转变，重新调整为一种更加全面的观念，即个人和公共福祉、生态平衡、对后代负责以及关爱地球——我们唯一的家园。我们还需面对另一现实：设计在引导我们走向新的繁荣过程中发挥着同样重要的作用，我们的主题展充分展现了这一潜力。

繁荣主题是本次主题展的总结篇，故事讲述了设计如何促进人与食物之间关系的改变、如何改变人的感受方式和信任方式，以及如何改变人与自然平衡相处的方式，同时又为我们发挥与生俱来的创新和精进动力提供了空间。人类文明就是在不断应对如何实现经济、社会和环境生态可持续发展的过程中不断走向繁荣的。

For a long time, we have measured our progress by economic growth, by accumulation of capital, by the salary we earn, by the distance we travel, by the things we consume, by the modes of our transport, by the size of our house, by the objects we possess. Many saw this material definition of wealth as equal to prosperity, and hence as the motivation to keep going in the same direction. And let's face it: design played a crucial role in the prevalence of this paradigm.

Today, a profound shift is happening, recalibrating this notion of prosperity towards a much more comprehensive idea of personal and mutual well-being, ecological balance, responsibility for future generations and care for the only planet we have. And let's face this as well, design is equally important in guiding us towards this direction, as exemplified in the main exhibition.

This pavilion acts as conclusive chapter to the story about the way design can help to change our relationship with what we eat, how we feel, how we trust and how we restore our balance with nature, while giving room to our innate drive to innovate and improve. Human civilization is constantly striving towards prosperity in the process of addressing how to achieve sustainable economic, social, and environmental development.

米兰垂直森林模型
Bosco Verticale Milan Physical Model

博埃里建筑设计事务所 | 2017
Stefano Boeri Architetti | 2017

尽管城镇化似乎是一种不可阻挡的趋势，但人们开始尝试构筑另一种城市环境，让其在城市中可以靠近树木、灌木和植物生活。"垂直森林"是一种展示建筑生物多样性的新型建筑原型，关注人类以及人与其他物种间的关系。世界首例垂直森林建于意大利米兰市中心，两座"树塔"分别高80米和112米，根据日照条件，其中种植了各类植物，包括800棵树，5 000株灌木丛，以及15 000株其他植物。这相当于种植了2万平方米的林地。本项目成功地在城市中心创造了丰富的植物多样性，探索了在人造城市环境中融入动植物物种的可能性，并促进了城市生态多样性分布点的发展。

While in many places urbanization still seems to be an unstoppable trend, people are exploring an alternative urban environment that allows them to live close to trees, shrubs and plants within the city, hereby supporting biodiversity. The Bosco Verticale is the prototype building for a new format of residential architecture, which focuses not only on human beings but also on the relationship between humans and other living species. The first Bosco Verticale counts two residential towers of 80 m and 112 m of height, realized in the center of Milan, it hosts 800 trees, a wide range of 5000 shrubs and 15,000 plants, distributed according to the exposure of the façades to sunlight. The Bosco Verticale equals, in amount of trees, an area of 20,000 m^2 of woodland. Such diversity and typology of plant species within the urban center works as a point of reference and a tool for urban policies directed to the inclusion of plant and animal species inside the man-made urban context, promoting the development of different urban biodiversity dissemination sites.

甜甜圈经济
Doughnut Economy

甜甜圈经济行动实验室 | 2016
Doughnut Economy Action Lab | 2016

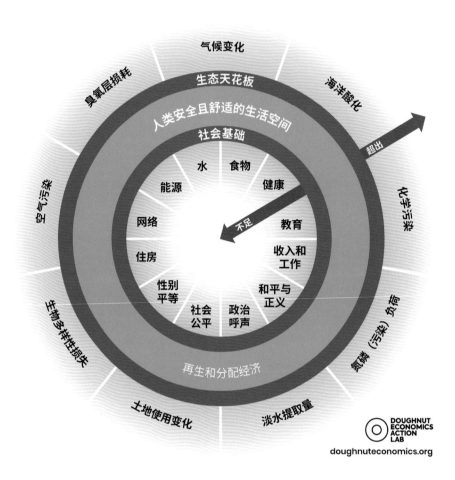

进入21世纪，气候变化、资源枯竭等需要人类共同应对的全球性危机日益加深，客观上要求人类对有限地球资源的利用控制在其承载力范围之内。因此，亟须制定出一套能够获得国际广泛认可的多维度可持续发展评价体系，以便于厘定出一条能够维持地球系统的自我再生，而又不危害后代生存的安全界限。"甜甜圈"理论，首次将社会经济系统结合到九项生物物理过程的行星边界框架中，使得区域/国家的可持续发展评估结果成为一个包含两个边界、三个区域的简化可视化模型。社会经济系统的提出还将经济增长的目标从"唯GDP论"转向多指标的共同发展。

In the 21st century, climate change, resource depletion and other global crises that require a common response from mankind are deepening, objectively requiring mankind to control the use of limited earth resources within its carrying capacity. Therefore, there is an urgent need to develop a multi-dimensional sustainable development evaluation system that can be widely recognized internationally in order to define a safe boundary that can maintain the self-regeneration of the Earth system without jeopardizing the survival of future generations. The proposed socio-economic system also shifts the goal of economic growth from a "GDP-only" approach to a multi-indicator approach to development.

无尽之形
Endless Form

张周捷 | 无尽之形实业（上海）有限公司 | 2018
Zhoujie Zhang | Endless Form Industrial (Shanghai) Ltd. | 2018

过去十年，中国家具行业经历了一个高速发展期。伴随着经济和科技的发展，传统的家具设计和制作方式已经不能满足人们对于居家环境个性化的追求。无尽之形（Endless Form®）是由张周捷数字实验室于2018年创立的一个具有开创性的家具品牌。这个以多样性和个性化著称的品牌，结合了创始人八年数字艺术和设计的经验，聚焦于视野前卫、高端的用户而推出了量产化和手工定制的数字家居产品。通过设计和科技的力量，实验室将塑造并推动数字时代的新生活图景。

Over the past decade, China's furniture industry has experienced rapid growth. As economic and technological advancements progress, traditional furniture design and production methods no longer meet the demand for personalized home environments. Endless Form®, founded by Zhang Zhoujie Digital Lab in 2018, is a pioneering furniture brand renowned for its diversity and personalization. Combining the founder's eight years of experience in digital art and design, this brand offers both mass-produced and custom-made digital home products for a high-end, avant-garde audience. Through design and technology, the lab shapes and drives a new vision of living in the digital age.

四百盒子社区
400 Box Community

青山周平 | 四百盒子科技有限公司 | 2016
Shuhei Aoyama | 400 Box Technology Quanzhou Co., Ltd. | 2016

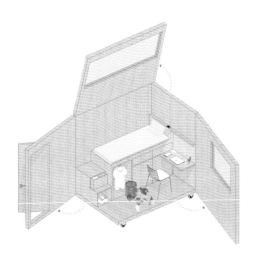

本项目为个体时代的城市年轻人构想了一种新型共享社区。本项目利用城市中空置的大楼,把过去用墙壁划分的房间打开,个人空间则被简化为5平方米的可移动盒子。厨房、卫生间、洗衣间——生活功能在盒子外集中布置,盒子间的区域可以自由布置,变成公用的影音室、书房。盒子内是私人领域,周围则是街坊邻里。这个方案的基本构想是把个人的私有空间缩至最小,腾出宽阔而丰富的公共空间,整体打造成一个迷你城市,或是家一般的共享社区。

This project envisions a new type of shared living environment for the urban youth of the individual era. By repurposing vacant urban buildings, this project opens up rooms that were previously divided by walls, reducing individual spaces to minimalist 5 m² mobile boxes. Essential living functions such as kitchens, bathrooms, and laundry facilities are consolidated outside the boxes, while the box interiors serve as personal spaces. The surrounding areas are designed to be flexible, transforming into shared spaces like media rooms and study areas. The core idea is to minimize private areas to maximize expansive and rich public spaces, creating a mini-city or home-like shared community.

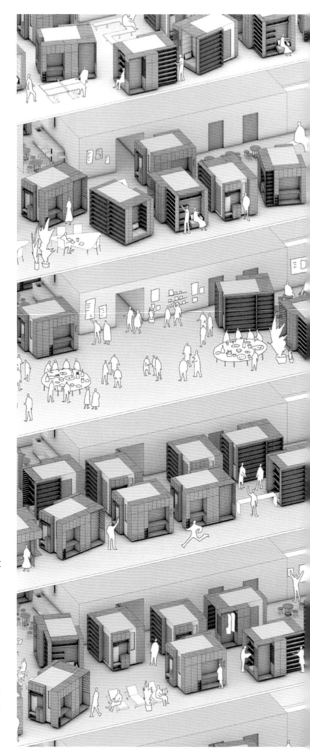

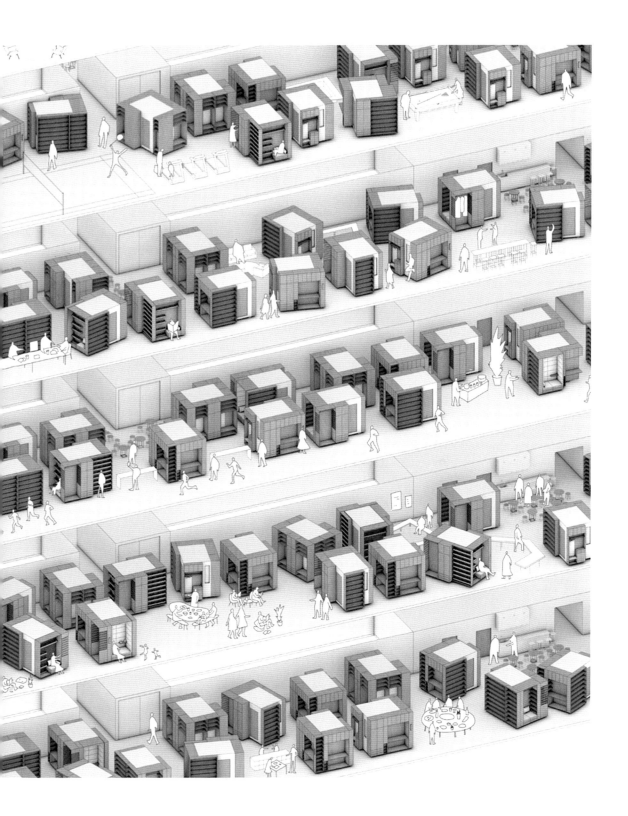

为了设计师的电路生产
Electronic Circuits for Designers

Fablab O Shanghai 数制工坊 | 2022
Fablab O Shanghai | 2022

Fablab O Shanghai数制工坊是同济大学设计创意学院2013年成立的一个创客工坊，通过提供数字制造工具，让用户能够"制造（几乎）任何东西"。它是中国大陆第一个全制式的微装配实验室，旨在为个体用户提供那些通常在大规模工业化生产当中应用的技术，让用户能够体验"个人制造"。设计师电子电路课程是该实验室提供的一门课程，旨在揭示电路板设计和制造的奥秘，并将制造过程从工厂带到工作室。在为期九周的课程中，学生将体验涉及设计、原型制作、制造、编程和测试复杂电子电路的工作流程，并将各种系统集成到一个真实的产品中。

Fablab O Shanghai, a makerspace of the College of Design and Innovation at Tongji University, offers digital fabrication tools enabling users to "make (almost) anything". As a member of the global Fab Lab network, it provides technologies typically used in large-scale industrial production, allowing users to experience "personal manufacturing". The "Electronic Circuits for Designers" course, offered by Fablab Shanghai, unveils the secrets of circuit board design and manufacturing, bringing the production process from factories into the studio. Over the course of nine weeks, students engage in the entire workflow of designing, prototyping, manufacturing, programming, and testing complex electronic circuits, integrating various systems into a real product.

BDD，与"繁荣"相生相成
BDD, Prosper with Society and Economy

设计驱动型品牌观察 | 博观必达（深圳）网络科技有限公司 | 2018
BDDWATCH | BOGUANBIDA（Shenzhen）Network Technology Co., Ltd. | 2018

设计驱动型品牌的高速发展是经济、社会"繁荣"的典型映照。一方面，设计驱动型品牌诞生于社会、经济与文化繁荣与稳定的大环境中，另一方面，设计驱动型品牌运用"以用户体验为核心"的设计思维洞察产业、行业、大众的潜在需求，以设计深度融合科技与商业，通过创造富有生命力、极致体验的优良设计产品与服务，极大地满足大众消费与经济发展的需求，真实有效地促进了社会的多元繁荣。

The rapid development of design-driven brands reflects the prosperity of both the economy and society. On one hand, these brands emerge within a context of social, economic, and cultural stability and growth. On the other hand, they leverage a "user experience-centered" design approach to explore potential needs in industries, sectors, and among the public. By integrating design deeply with technology and business, design-driven brands create vibrant and exceptional products and services that greatly satisfy consumer demands and economic development, thereby effectively fostering diverse societal prosperity.

BDDWATCH

BDD，与「繁荣」相生相成

BDDWATCH是一个专注于研究、传播、推动、赋能设计驱动型品牌（Brand Driven by Design，简称 BDD）理念的内容生产架构，由童慧明教授于 2018 年初发起创立。

通过近5年持续对BDD趋势及典型案例的创业成长研究，有三个显著的趋势呈现出来：首先，基于大众普遍产生更高层级的物质与精神需求，拉动消费向高品质生活体验的升级；其次，中国制造水平、投资环境和销售渠道日益完善和开放，为好设计品牌的创立和发展奠定了坚实基础；最后，伴随设计从物理化的"有形设计"向抽象逻辑化的"无形设计"思维嬗变，从企业中下层的"专业、执行性设计"向"系统、战略型设计"顶层的跃升与组织架构再造，设计产业以及商业社会对设计价值、系统设计逻辑维度边界的认知逐步突破。

BDD品牌的高速发展是经济、社会"繁荣"的典型映照。一方面，BDD品牌诞生于在社会、经济与文化繁荣与稳定的大环境，另一方面BDD品牌运用"以用户体验为核心"的设计思维洞察产业、行业、大众的潜在需求，以设计深度融合科技与商业，通过创造富有生命力、极致体验的优良设计产品与服务，极大地满足大众消费与经济发展的需求，真实有效地促进了社会的多元繁荣。

在本次世界设计之都大会上，BDDWATCH将展示8个BDD案例，既有深扎数字化领域、融合创意内容与技术创新、构建创意设计数字新基建的特赞，也有在大健康领域引领我国高端医疗设备创新方向、用人性化设计打造温暖诊疗体验的联影医疗。而在消费品领域，诸如以品牌矩阵登临全球婴童用品顶峰的好孩子、专注于打造极致儿童竞技运动体验的 bike 8、用新生活工具引领新生活方式的小猴工具、让美味更简单的TOKIT、让养宠更愉悦的pidan以及提出更适宜中国家庭健身解决方案的梵品，都在各自聚焦的行业里用持之以恒的设计创新，在表达独特价值主张的同时，用美好的产品改变世界。

 特赞
创意设计的数字新基建

bike 8
陪伴儿童成长的运动体验

HOTO 小猴工具
极品生活工具打造者

pidan pidan
享受生命陪伴的乐趣

联影医疗
大国重器的人性化之美

 梵品
重新定义家庭健身

TOKIT
用温度定义烹饪态度

NICE2035 未来生活原型街
NICE2035 Living Line

同济大学设计创意学院 | 2018
College of Design and Innovation, Tongji University | 2018

NICE2035未来生活原型街（简称NICE2035）的初衷源于普通人基于日常生活的智慧是社会创新的源头，也是重要的设计灵感财富。社会生活的组织，可能对空间"活化"起到了更为关键的作用。同济大学设计创意学院联手四平路街道和著名企业及创新工作室，希望在鞍山五村这条200米不到的工人新村的小巷里，提供一个合适的平台，让青年人回到社区，推动社区的改变和意义的创造。通过大学、社区与园区的融合，共同探索未来生活的可能性。

The NICE2035 Living Line project (NICE2035) originated from the idea that everyday wisdom of ordinary people can be a source of social innovation and significant design inspiration. The organization of social life plays a crucial role in activating spaces. In collaboration with the Siping Road Subdistrict, prominent companies, and innovation studios, Tongji University's College of Design and Innovation aims to provide a platform for young people to engage with their community, fostering change and meaningful impact. This initiative explores future living possibilities through the integration of universities, communities, and innovation parks, with a focus on transforming a small alley in Anshan Wu Cun into a vibrant, interactive space.

旧里新厅：贵州西里弄微更新
Shared-Living Space: Micro Regeneration Projects in West Guizhou Lilong

梓耘斋建筑 | 2018
TM Studio | 2018

如同大多数的上海传统里弄，贵州西里弄社区有近百年历史，其中许多家庭内部空间狭小、设施破损、品质陈旧。随着城市社会变迁，社区内部的结构关系复杂化，导致了日常维护与修缮问题增多，居民自我更新能力减弱。本项目探讨如何利用有限的资源与环境，通过12个社区触媒点的微创新改造，为居民提供必要的空间与设施。通过营造1 800平方米的社区共享客厅，提升公共生活品质，加强场所归属，凝聚居民共识，从而带动社区朝向美好生活。

Like many traditional lilong neighborhoods in Shanghai, the West Guizhou Lilong community, with nearly a century of history, features small, dilapidated, and outdated family spaces. As urban social dynamics have evolved, the complex internal structures of the community have led to increased maintenance and repair issues, weakening residents' capacity for self-renewal. This project explores how limited resources and environments can be leveraged through 12 micro-innovation touchpoints to provide essential space and facilities for residents. By creating a 1,800 m² community shared living room, the project aims to enhance public life quality, strengthen a sense of place, and build community consensus, thereby fostering a transition towards a better living environment.

湖南路街道综合服务站
Hunan Road Subdistrict Public Service Station

袁烨，张子岳 | 热气建筑工作室 | 2022
Ye Yuan, Ziyue Zhang | Steamarchitecture | 2022

本项目立足于上海十五分钟生活圈与微更新计划，通过对既有环卫设施的改造，营造更为丰富的城市公共空间。设计通过平面功能配置的腾挪，使改善后的空间与现状结构、设备以及风貌区立面相协调。设计关照女性、儿童及宠物，并为环卫工人与管理人员提供了更舒适的办公与休息空间。重新开放后，市民将从入口花园通过公共走廊进入建筑，并可体验到内部天井庭院与更易使用的公厕设施。

This renovation focuses on Shanghai's 15-minute living circle and micro-regeneration plan by enhancing existing sanitation facilities to create richer urban public spaces. The design involves repositioning functional areas to harmonize the updated space with existing structures, equipment, and façade aesthetics. It caters to women, children, and pets, while providing more comfortable office and rest areas for sanitation workers and management staff. After reopening, citizens will enter the building through an entrance garden and public corridor, experiencing an internal courtyard and upgraded restroom facilities designed for better usability.

脑机比 IP 系列艺术作品
Brain-Machine-Ratio IP Series Artworks

脑机比工作室 | 2022
BMRLAB | 2022

BMR是Brain-Machine-Ratio（脑机比）的缩写，指人脑的感性思维与机器的理性思维之间的分工协作比例。其核心是人机共同进化，倡导人类与技术的共生发展。在人工智能与数字经济的时代背景下，脑机比工作室选择历史上的伟大人物与经典工具，从脑和机的两个维度进行拆解，创作出可爱的IP数字艺术品。脑机比探索纯真、艺术、科学，连接起过去、现在和未来。

BMR stands for Brain-Machine-Ratio, reflecting the division of labor between human intuitive thinking and machine rational thinking. Central to this concept is the idea of co-evolution between humans and technology, promoting symbiotic development. In the context of artificial intelligence and the digital economy, BMRLAB draws inspiration from historical figures and classic tools, deconstructing them from both brain and machine perspectives to create charming IP digital artworks. BMRLAB explores the themes of innocence, art, science, connects past, present, and future.

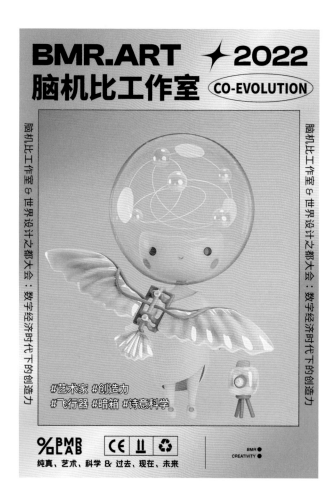

"奇鳞"机械屏
"Qilin" Mechanical Screen

吴宽 | 马努（上海）艺术设计有限公司 | 2022
Kuan Wu | Manu | 2022

本产品也可以被称为可动建筑表皮，是一种神奇的动态鳞片集群。每一片鳞片都有一个独立的电机精确地控制运动角度。几万个鳞片聚集在一起，播放动态媒体画面，具有强烈的"机械印象派"装饰风格，蔚为壮观。它适用于建筑与空间的主体形象墙，可吸引民众前来观赏。从技术上讲，它拥有超大电机集群、3~20分贝的超静音、机械与光电一体集成、低功耗、无光污染、不受断电干扰等优势，不仅可在物理层面实现"播放"，更通过独特的人与屏之间艺术情感互动，能衍生出独特的多维体验和艺术表现。

The "Qilin" Mechanical Screen, officially known as a movable architectural skin, is a mesmerizing array of dynamic scales. Each scale features an individual motor that precisely controls its movement angle. When combined, thousands of these scales create dynamic media displays with a striking "mechanical impressionism" style, producing a breathtaking effect. Ideal for architectural and spatial feature walls, this screen also serves as an eye-catching landmark. Technically advanced, it boasts large motor clusters, ultra-quiet operation (3~20 decibels), integrated mechanical and optoelectronic systems, low power consumption, no light pollution, and resistance to power interruptions. Beyond its physical display capabilities, it fosters unique artistic interactions between viewers and the screen, offering a multifaceted experience and artistic expression.

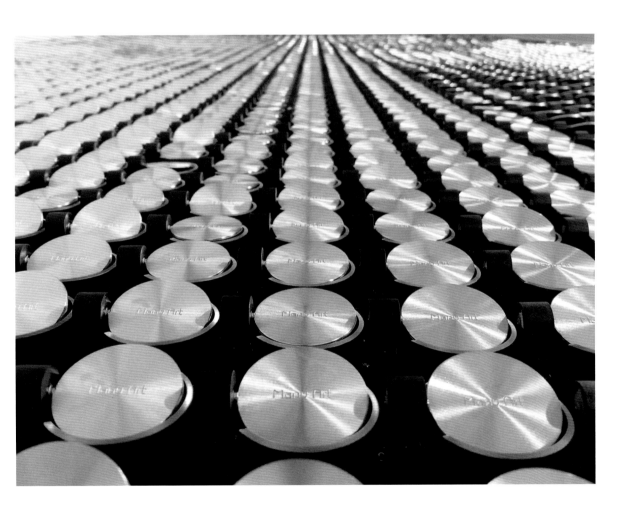

1923，幻镜上海
1923, Fantastic World Shanghai

张文绮 | 上海三联（集团）有限公司 | 2022
Vikki Zhang | Shanghai Sanlian (Group) Co. Ltd. | 2022

"海纳百川"是上海的城市精神，繁荣的上海从1923年开始。1923年，中华文化与西洋文化在开埠的上海交会，孕育出了开放融合的海派文化。刚刚接受西方先锋性思想的海派潮人，用独树一帜的视野，亲身实践他们想象中的未来。他们大胆创新，天马行空，追求一个理想的新境界。一个城市的繁荣，并不简单地由经济的指标来佐证，更是看这里的人们有多潮流和先锋，这里的思想有多开放和包容。

The spirit of Shanghai is embodied in the phrase "sea embraces all rivers", reflecting the city's openness and prosperity, which began in 1923. This year marks the convergence of Chinese and Western cultures in Shanghai, giving rise to the cosmopolitan Haipai culture. Embracing avant-garde Western ideas, the Haipai people of the time boldly innovated and envisioned a future with unique perspectives. The prosperity of a city is not merely measured by economic indicators but also by its trends and avant-garde spirit, and the openness and inclusivity of its ideas.

戴森设计大奖
The James Dyson Award

自 2005 年起
Since 2005

戴森设计大奖是一项国际性的设计与工程大奖，由戴森创始人詹姆斯·戴森创立于2005年，旨在嘉奖、鼓励和启发下一代工程师和发明家。它面向所有工程专业或设计专业的大学生和新近毕业生开放征集，宗旨为"设计一个能够解决问题的方案"。自2016年进入中国的七年来，戴森设计大奖见证了成千上万中国青年发明人才和成长。开发一项产品或技术是一个漫长而充满挑战的过程，戴森设计大奖将为有意愿投身这一领域的年轻才俊提供平台，为脱颖而出的新生代发明人才提供支持。

The James Dyson Award, founded in 2005, is an international design award that forms part of a wider commitment by James Dyson, to inspire and encourage the next generation of engineers and inventors. The Award challenges entrepreneurial undergraduates and recent graduates of engineering and design to "design something that solves a problem". From 2016, the Award has challenged thousands of Chinese design and engineering university students. It takes time, energy and pains to develop a new product or technology. The James Dyson Award provides a platform to support the young inventors who are willing to commercialize their designs.

HOPES：眼压检测器
HOPES: Home Eye Pressure E-Skin Sensor

鱼珂露，李思，大卫·李 | 新加坡国立大学 | 2021
Kelu Yu, Si Li, David Lee | National University of Singapore | 2021

眼压数据作为临床医生评估青光眼的唯一指标，完成定期监测、收集数据对青光眼的长期治疗非常重要。然而，在家自行测量眼压通常不够准确，而更准确的Goldmann压平眼压计测量法仍需在临床进行。而HOPES是一种可以居家无痛检查眼压的可穿戴生物医疗器械，成本低廉，简便易用。它使用高密度压力传感器阵列来捕捉角膜的高分辨率压力梯度图。通过机器学习算法，实现了简单、快速、准确的眼压测量。

HOPES, (which stands for Home Eye Pressure E-skin Sensor) is a wearable biomedical device for pain-free, low cost, athome IOP (Intraocular Pressure) testing. Powered by patent pending sensor technology and artificial intelligence, HOPES is a convenient device for users to frequently self-monitor IOP. As the only indicator for glaucoma, regular IOP monitoring is a critical tool in helping clinicians to determine long-term treatment plans and goals of glaucoma. Self-measurement of IOP at home is often inaccurate, while the more accurate Goldmann Applanation Tonometry still needs to be performed clinically.

照片由戴森设计大奖提供
Photo credit: The James Dyson Award

照片由戴森设计大奖提供
Photo credit: The James Dyson Award

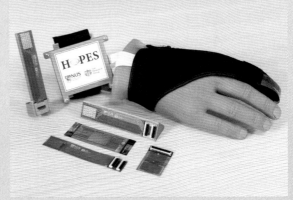

REACT 快速止血系统
REACT: A Bleed Control System

约瑟夫·本特利 | 英国拉夫堡大学 | 2021
Joseph Bentley | Loughborough University | 2021

REACT系统使用一种快速充气的止血装置，可插入刺伤造成的伤口。这个止血装置可自动充气，直接对伤口施加内部压力，能更快地控制出血。急救人员可将本系统中医用级别的硅胶球囊止血装置插入伤员伤口。在急救现场警察通常是最早到达的受过训练的应急人员，但他们没有快速且易于使用的工具来防止伤员失血过多。治疗刺伤的建议是不要拔出被刺入的物体。这是因为物体既对伤口施加内部压力，同时也填充了腔体，可防止内部出血。REACT的设计就是基于这一原理。

The REACT system uses a rapid, inflatable Tamponade device that is inserted into the stab wound. The automated inflation of this Tamponade provides internal pressure direct to the bleeding site, controlling bleeding faster than current methods. The implantable medical-grade silicon Balloon Tamponade is inserted into the wound tract by a first responder. The Police are often the first trained responders at the scene, but they do not have the rapid and accessible tools required to prevent blood loss. The advice for treating stab wounds is to never remove the impaled object. This is because the object is applying internal pressure to the wound site whilst also filling the cavity, preventing internal bleeding. That's what REACT solves.

照片由戴森设计大奖提供
Photo credit: The James Dyson Award

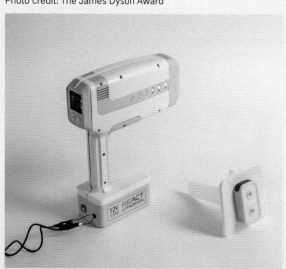

塑料扫描器
Plastic Scanner

杰里·德沃斯 | 荷兰代尔夫特理工大学 | 2021
Jerry de Vos | Delft University of Technology | 2021

塑料扫描器使用离散近红外光谱技术来检测塑料的类型，这是一种新型且低成本的传统红外光谱方法。它可以告诉用户塑料制品的材料成分。此外，它是完全开源的硬件，任何人都可以组装一个电路板并嵌入电子元件，以便专业人士提供反馈和改进意见。塑料是耐用的商品，通过特定的技术应用，可以广泛回收并转化为新产品。这个解决方案的灵感来自于一个大型本地工厂使用红外反射来对塑料进行分类的成功案例。

Plastic Scanner uses near-infrared spectroscopy to detect types of plastic—a new and low-cost approach to traditional infrared spectroscopy. It tells users what materials it's made from. It's also fully open-source hardware, and anyone can assemble a board and embed electronics into it, allowing professionals to provide feedback and improvements. Plastics are durable goods that, when applied with a specific technology, can be widely recycled and transformed into new products. The solution was inspired by the successful case of a large local factory using infrared reflection to sort plastics.

照片由戴森设计大奖提供
Photo credit: The James Dyson Award

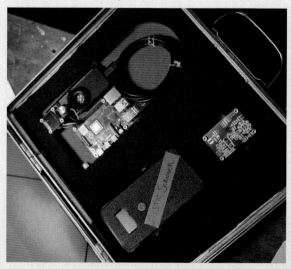

蓝盒子：家用乳腺癌筛查设备
The Blue Box: A Home-use Breast Cancer Biomedical Detection Device

朱迪特·吉罗·贝内特 | 西班牙巴塞罗那大学 | 2020
Judit Giró Benet | University of Barcelona | 2020

农作物废料太阳能发电系统
AuREUS System

卡维·埃伦·麦格 | 菲律宾玛普阿大学 | 2020
Carvey Ehren Maigue | Mapua University | 2020

目前乳腺癌的筛查流程要求妇女到医院或医疗机构进行侵入性检查，疼痛感强且费用高昂。而The Blue Box 是一款家用的乳腺癌生物医学检测设备，它利用尿液样本和人工智能算法来检测乳腺癌早期症状。本设备是非侵入性的、无痛的、无辐射的、低成本的，让女性在家就可以检测乳腺癌，掌控自己的健康。本设备对尿液样本进行化学分析后，将样本数据上传至云端服务器，基于人工智能的算法对尿液中的特定代谢物做出反应，为用户提供诊断参考。

The current breast cancer screening process requires women to go to a hospital or medical facility for invasive, painful and expensive tests. The Blue Box is a home-use breast cancer biomedical detection device that uses urine samples and Artificial Intelligence algorithms to detect early breast cancer symptoms. The device is non-invasive, painless, radiation-free, and low-cost, allowing women to detect breast cancer at home and take control of their health. After chemical analysis of the urine sample, the Blue Box uploads the sample data to a cloud server, and the artificial intelligence-based algorithm responds to specific metabolites in the urine to provide users with a diagnosis.

AuREUS系统是用于墙壁和窗户的一种进化型技术，它使用从再生农作物废料中合成的技术来吸收阳光中的散射紫外线，并将其转化为清洁可再生的电能。相较于太阳能电池板，本系统能更有效地将太阳能转化为可再生能源，即使没有直接的阳光照射。测试表明，其光电转换效率达到48%，而传统光伏发电的效率仅为10%~25%。

The AuREUS system is an evolution for walls/windows, and uses technology synthesized from upcycled crop waste to absorb stray UV light from sunlight and convert it to clean renewable electricity. AuREUS promises to convert more solar energy into renewable energy than solar panels can, even when there is no direct sunlight. Tests show that its photoelectric conversion efficiency reaches 48%, while that of traditional photovoltaic power generation is only 10~25%.

照片由戴森设计大奖提供
Photo credit: The James Dyson Award

照片由戴森设计大奖提供
Photo credit: The James Dyson Award

未来出行
Next Mobility

刘震元, 刘胧, 樊中, 赵华森, 王琦 | 同济大学设计创意学院 | 2021
Zhenyuan Liu, Long Liu, Zhong Fan, Huasen Zhao, Qi Wang | College of Design and Innovation, Tongji University | 2021

出行设计涉及对出行相关的产品、服务、体验、交互、系统的环境的创新设计和全新定义，但在此之前，需要通过对技术、社会和经济的深层讨论、反思、批判，去寻找未来出行概念的灵感。以下十个作品来自同济大学设计创意学院的前沿产学研联合课程设计成果，包括南极离站科考车、火星玩家、同济-广汽"流动的舞台"系列、同济-阿斯顿·马丁创意实验室的课题，以及梅赛德斯-奔驰激发的智能出行体验等毕业设计项目。本系列作品通过设计作品发表与学生路演，将未来出行这一主题从研究和设计对象，拓展为共同探究的话题，并通过设计连接更真实、更多元和更广阔的世界与挑战。

The Exhibition Next Mobility explores the future of transportation through innovative design of products, services, experiences, interactions, and systems. This exhibition features cutting-edge academic and industry collaboration projects from Tongji University's design programs. It includes concepts such as the Cross-Anta Vehicle, Mars Player, Tongji-GAC "Mobile Stage" series, and Tongji-Aston Martin Creative Lab, along with a smart mobility experience design inspired by Mercedes-Benz. Combined with students' roadshow, the exhibition extends the theme of future mobility into a collaborative exploration, connecting design with a more tangible, diverse, and expansive understanding of global challenges.

数据乌托邦
DATAPIA

刘臻莉 , 蔡学荣 , 吴滟芸 , 潘瑞琨 | 2021
Zhenli Liu, Xuerong Cai, Yanyun Wu, Ruikun Pan | 2021

数据乌托邦是全民参与数据流动的新世界。本产品作为维系数据系统健康运转的公众平台，致力于汇集数据所有者贡献的信息，使其流向最有价值的应用端。通过个性化移动数据采集器，每个人都能在多元场景下获取并存储数据——从城市交通流动到恋人相拥的温暖，所有数据都能通过本产品转化为有价值的资源。借助区块链等技术，本产品保障每位用户身份信息的隐秘性与数据的可追踪性。尽管纯粹美好的数据乌托邦也许不会到来，但设计师们坚信，每个公民的自主参与都可能让数据时代更趋完美。

DATAPIA represents a new world of data flow driven by universal participation. This public platform aims to maintain the health of data systems by aggregating information from data owners and directing it to valuable applications. Using personalized mobile data collectors, individuals can gather and store data in diverse contexts—from city traffic to personal moments of warmth. DATAPIA transforms this data into valuable resources, leveraging technologies like blockchain to ensure user privacy and data traceability. While a pure utopia may remain elusive, the design team believes that citizen involvement can enhance the data era's effectiveness.

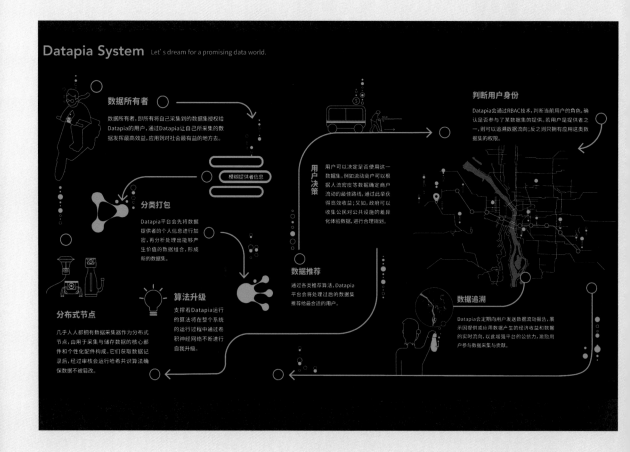

菌基互黏出行
Mobility as Physaru

陈鸣惊, 陈宁子, 肖雨欣, 熊天琦 | 2021
Mingjing Chen, Ningzi Chen, Yuxin Xiao, Tianqi Xiong | 2021

人类作为生物无法脱离为自身考虑的视角，无法站在自然的角度思考做到真正的可持续，因此我们提出了去人类中心主义。在未来，我们需要一个第三方的角色连接人和自然——一种经生物技术改造为生物计算机的黏菌。它遍布自然，收集自然中的信息与人类的移动信息，站在整个系统的视角，来代替人类计算、规划与决策。当人类出行时，与地面接触的转译器官向黏菌网络传达出行需求，并获得反馈；人乘坐的载具由生物材料构成，并由黏菌驱动运转。

Human beings often struggle to adopt a truly sustainable perspective. To address this, the concept of "de-anthropocentrism" is introduced. It proposes a third-party role—an organism transformed into a biological computer through biotechnology, such as slime molds. These organisms, which are widespread in nature, collect information from both the environment and human movements, offering a holistic system perspective to replace human-centric computation, planning, and decision-making. When people travel, their movement needs are communicated to the slime mold network through contact sensors, which then provide feedback. Vehicles used by people are constructed from biological materials and operated by the slime mold network.

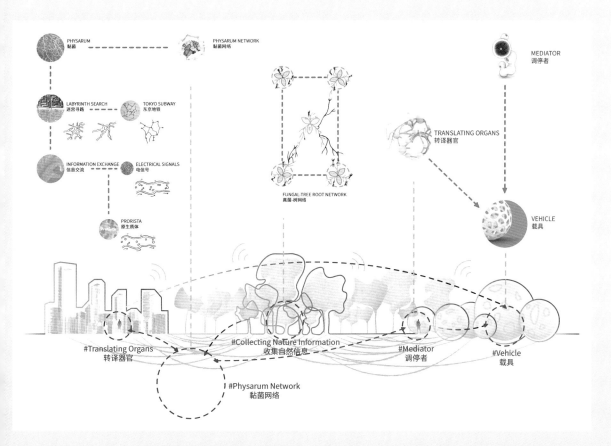

且行且舞
POSA

巩珑钰，李晋，侍建宇，陈宣宇 | 2021
Longyu Gong, Jin Li, Jianyu Shi, Xuanyu Chen | 2021

高度发达的数字技术和无人驾驶技术让我们看到了一个解放人类驾驶角色的近未来出行图景。然而解放的身体又会被代入怎样的体验呢？我们真的想被装在小盒子里从一个地方运输到另一个地方吗？我们真的想终日沉浸在虚拟世界带来的数字幻象之中吗？出行本该多彩，本作品希望以流动的姿态为出行赋予无限的精彩。且行且舞，让我们拥抱生动真实的出行体验和多样活力的未来。

Advanced digital and autonomous driving technologies offer a near-future vision of travel where human drivers are liberated from their roles. However, what kind of experience will this liberation bring? Do we truly want to be confined in small boxes, transported from one place to another? Do we wish to be perpetually lost in the digital illusions of virtual worlds? Travel is supposed to be vibrant and dynamic. This work aims to infuse the journey with the beauty of motion and diverse vitality. POSA envisions a future where travel is an engaging, lively experience, embracing the richness and energy of real-world movement.

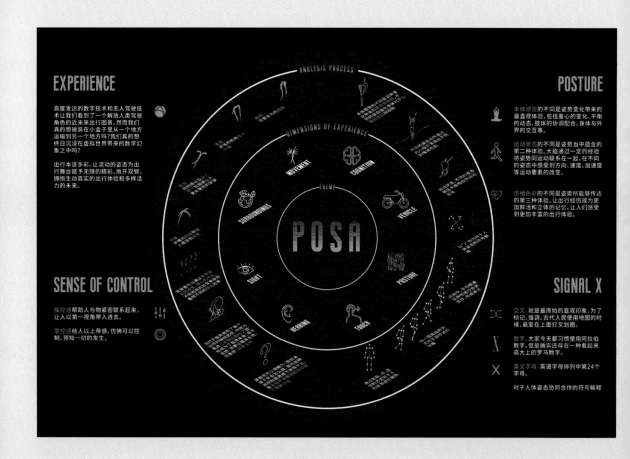

自然语者
DIALOG ATTORNEY

郑茜予，魏倩洁，高世涵，陈安 | 2021
Qianyu Zheng, Qianjie Wei, Shihan Gao, An Chen | 2021

我们生活在经验主义者的噩梦中：一个真实的世界，存在于我们的知觉范围之外。当我们进入自然中，便被卷入密布的通讯网和关系网中，无论是否意识到，我们都改变了这些网络，而我们本身就该是一份子。我们向自然系统派去"人造突触"，以自然的方式融入自然，去链接，去对话，去呈现，感知并且传递信息。所有的生命体与非生命体分享着世界，我们与万物的感知不期而遇。自然在说什么？树、石、土……它们会像我们一样观看，将方格卡片的影像呈现在它们那疏管式的心灵中吗？它们的内心也会产生对光明与黑暗的体验，并由神经加工成意愿、偏好和意义吗？本作品试图寻找答案。

We find ourselves in the nightmare of empiricists: a real world that exists beyond our perceptual range. When we immerse ourselves in nature, we are entangled in complex communication and relational networks, altering them whether we are aware of it or not. We should be an integral part of these systems. By deploying "artificial synapses" we aim to naturally integrate with nature, facilitating connections, dialogues, and information transmission. All life forms and non-living entities share the world, and our perceptions unexpectedly intersect with theirs. What does nature communicate? Do trees, stones, soil, etc., perceive the world and process images like we do? Do they experience light and darkness, processing them into intentions, preferences, and meanings? This work seeks answers to these questions.

比特之息
The Breath of Bits

袁佳琳，瞿欣雨，何雨聪，蔡同 | 2021
Jialin Yuan, Xinyu Qu, Yucong He, Tong Cai | 2021

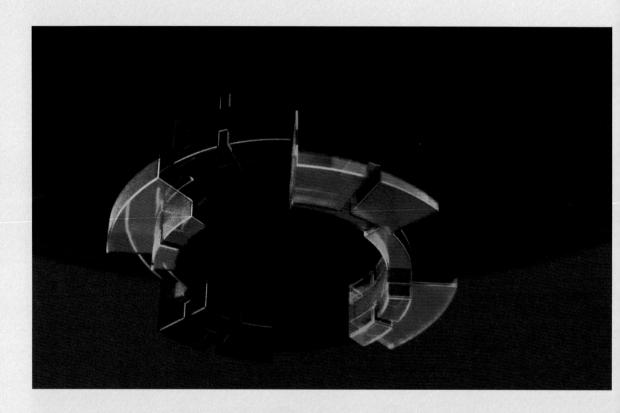

我们正以超音速进入比特世界，但如何感知这个未知领域的气息？扰动的信息是比特世界的元，恰如同中国八卦中的"爻"，"0"和"1"、阴与阳，构成了复杂世界的计算法则。阴阳交叠，原子世界的万物被编译成代码；0和1的排列能否解译比特世界的一切？听觉，作为视觉范式下被忽视的重要感知途径，极具潜力。声音特有的符号性与波动性有望实现信息的强压缩。未来人工智能操控的出行系统将削弱人类原生感官的认知，本作品将为人类提供倾听比特之息的可能，拓展全知全能的边界。正所谓："比特悬音，坎耳为闻；且听声爻，流动乾坤。"

We are supersonically entering the realm of bits, but how can we sense the essence of this unknown domain? Disruptive information forms the essence of the Bit World, akin to the hexagrams in the Chinese *I Ching,* where "0" and "1", *Yin* and *Yang*, create the computational rules of a complex world. As *Yin* and *Yang* overlap, the physical world's matter is compiled into code; can everything in the Bit World be deciphered from the arrangement of 0 and 1? Auditory perception, often overlooked in the visual paradigms, holds significant potential. The unique symbolism and fluctuation of sound offer promising information compression. As future AI-driven mobility systems diminish human sensory cognition, this work aims to provide a means for humanity to listen to the "breath of bits", expanding the boundaries of omniscience.

"游·乐·场"：2050 年的未来出行
"Playground": the Future of Travel in 2050

闻一然 | 同济 - 阿斯顿·马丁创意实验室 | 2021
Yiran Wen | Tongji-Aston Martin Creative Lab | 2021

本作品将 2050 年的未来城乡定义为一个大型"游·乐·场"，以未来出行为聚焦点，以设计虚构为思考方式，为美好的明天而设计，探索未来可能的载具 / 出行服务体系，提升全民幸福感，构建未来城市 / 乡村"游·乐·场"。

This project envisions the future of urban and rural areas in 2050 as a vast "Playground", with a focus on future mobility. Using Design Fiction as a thinking tool, it explores potential vehicles and travel service systems to enhance public happiness and create a future city and countryside that resemble a dynamic and engaging "Playground". The aim is to design for a better tomorrow by reimagining travel and its role in future urban and rural environments.

智能时代下的出行体验设计
Mercedes-Benz EQX

王惜轿 | 2021
Xiqiao Wang | 2021

汽车是移动的信息源，随着自动驾驶的发展和屏幕世界的扩张，曾经多方位的信息源将不断整合、统一，最终构筑为半包裹的"信息"空间。在未来的驾乘过程中，人们不再分散感知多信息源，而是感知整个信息空间。在信息空间中，结合数字虚拟技术，可在旅程中呈现不同自然景观或个性化艺术空间，为用户创造虚拟沉浸体验。

Automobiles are evolving into mobile information sources. With the advancement of autonomous driving and the expansion of screen-based environments, the formerly diverse sources of information will increasingly consolidate into a semi-enclosed "information" space. In the future, passengers will not experience scattered information sources but rather perceive a unified information space. By integrating digital virtual technologies, this space can present various natural landscapes or personalized artistic environments during the journey, offering users immersive virtual experiences.

南极离站科考车
Cross-Anta

黄佳捷 | 2021
Jiajie Huang | 2021

SCENARIO
生存车作为先遣车辆，科考设备车自动跟车，
离站科考正式开始。

从古至今，好奇心和创造力驱使着人类不断进步和发展。而今人类的探索足迹虽已深至海底远至宇宙，地球上却尚存有一片充满未知的大陆——南极。人类虽已在南极大陆建成多个科考站，并以此为基础在站周边开展工作，但人类的足迹尚未覆盖至南极大部分地区。在未来，技术的发展将使得南极"离站科考"成为可能。本作品正是一个帮助实现"离站科考"的交通工具。

Curiosity and creativity have driven human progress and development throughout history. From exploring the depths of the oceans to venturing into space, human beings have continually pushed boundaries. Yet, there is still a most enigmatic continents remains largely unexplored—Antarctica. While multiple research stations have been established on this continent, many areas still remain untouched. As technology advances, the possibility of off-station exploration in Antarctica becomes feasible. This project focuses on designing a vehicle that facilitates "off-station scientific exploration", enhancing our ability to explore the vast, uncharted regions of the Antarctic.

火星玩家
Mars Player

2018 级工业设计 | 2019
Class of 2018 Industrial Design | 2019

在一个全新的星球，我们应该如何应对陌生而充满挑战的生存环境，如何建立起新的社会和文明，如何构建更公平的秩序、更可持续的发展？借助"火星"这个对人类既熟悉又陌生的设计背景，本课题希望启发学生对未来探索做有意义的设计尝试。本课题要求学生回答以下问题：如何基于数据抓取构建认知未来的信息链条？如何梳理行为活动与功能造型间的因果关系？如何达成意料之外且情理之中的设计成果？

In the context of a new planet, how should we address the unfamiliar and challenging survival environment? How can we establish new societies and civilizations, and build a fairer order and more sustainable development? Using Mars as a design background that is both familiar and alien to humanity, this project aims to inspire students to make meaningful design attempts for future exploration. Students are tasked with addressing the following questions: How can we construct a cognitive information chain for the future based on data capture? How do we relate behavioral activities to functional forms? How can we achieve design outcomes that are both surprising and sensible?

无线充电太阳能光伏三轮车
Wireless Charging Tricycle with Solar Charging System

苏运升，彭震宇，康海星，库迪 | 同济大学上海国际设计创新研究院 | 2020

Yunsheng Su, Zhenyu Peng, Haixing Kang, Di Ku | Shanghai International College of Design and Innovation, Tongji University | 2020

本作品运用太阳能光伏清洁能源和无线充电技术，以求推动轻出行的发展。同时，本作品也具备了L4级的自动驾驶功能。这套解决方案配合区块链技术和C2C商业模式，不仅能用于客货营运与旅游观光，还可以打开个性化潮车市场，打造属于年轻群体的潮玩社交圈。

This project integrates solar photovoltaic clean energy with wireless charging technology to advance lightweight mobility. It also features Level 4 autonomous driving capabilities. The solution, combined with blockchain technology and a C2C business model, is designed for passenger and freight operations as well as tourism, while also tapping into the personalized and trendy vehicle market. It aims to create a social sphere for the younger generation.

展览设计
Exhibition Design

为了最大限度地减少碳足迹，展陈设计采用了装配式策略。我们与阿旺特合作设计了可以反复使用的装配式展墙系统，用1 000个红酒箱，按照主题展的6个主题，堆叠出各种不同的形态作为展览的主要陈列载体。后来这些红酒箱被反复利用，参加了无数的展览，也寓意WDCC的设计精神的传播。

In order to minimize carbon footprint, the exhibition design adopts an assembly strategy. We collaborated with AVARTE to design a reusable prefabricated exhibition wall system, using 1,000 wine boxes stacked in various forms according to the six themes of the exhibition as the main display carriers. Later on, these red wine boxes were repeatedly used and participated in numerous exhibitions, symbolizing the dissemination of WDCC's design spirit.

主创构思：娄永琪

设计执行：郭泠

Creative Concept: Yongqi Lou

Design Execution: Ling Guo

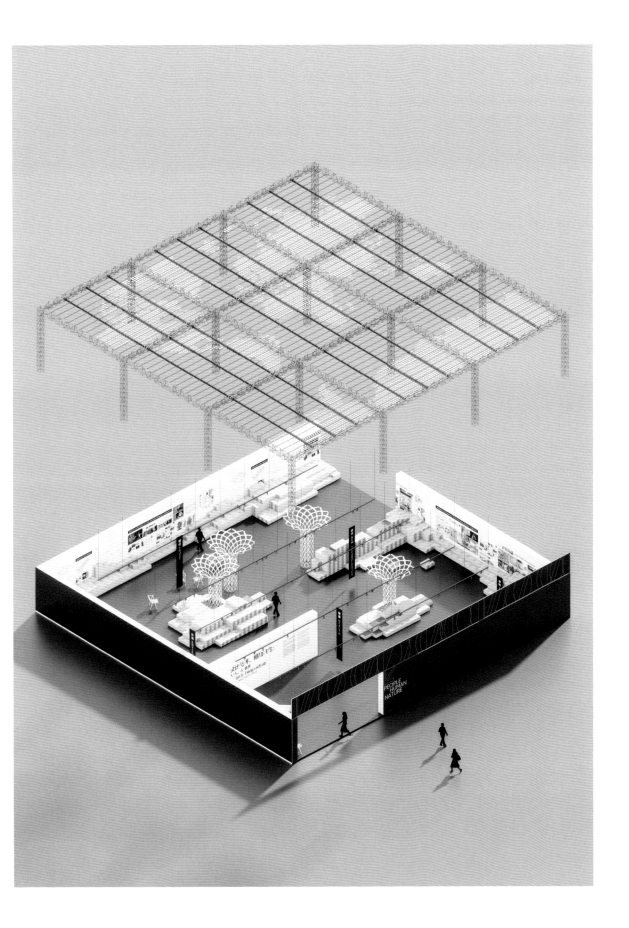

主题展"繁荣"馆
章明教授设计
Theme Exhibition "Prosperity" Pavilion
Designed by Prof. Ming Zhang

主题展"繁荣"馆
章明教授设计
Theme Exhibition "Prosperity" Pavilion
Designed by Prof. Ming Zhang

Exhibition Design

展览实景
Exhibition Scenes

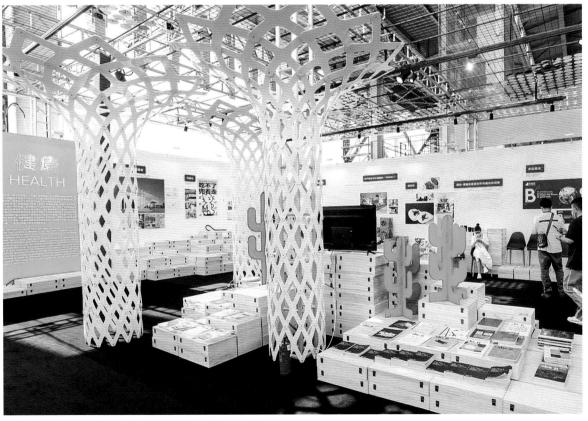

Exhibition Scenes

Exhibition Scenes

展览实景

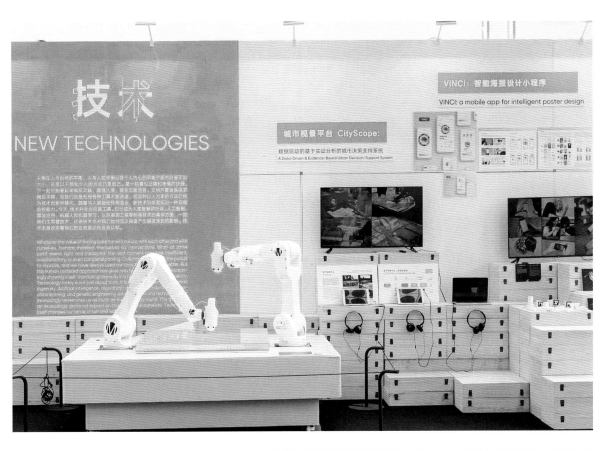

Exhibition Scenes

展览实景

Exhibition Scenes

Exhibition Scenes

Exhibition Scenes

工作坊
Workshops

循环设计：
红酒盒回收利用计划

合作伙伴: 同济大学设计创意学院, 哥德芬, 万华化学

特别支持: 艾伦·麦克阿瑟基金会

学术主持: 杨皓, 莫娇

活动简介: 我们的经济很大程度上依赖大量快速消费品而繁荣。这些为短暂使用而生产的物品，能否有更长的生命周期，在其所蕴含的营养元素回归自然系统之前，得到最大化的利用，以产生价值？红酒盒群岛和塑料竹林营造了主题馆空间，展览结束后如何被再使用、再制造和再回收？工作坊将以循环经济、技术创新、人本洞察、创意想象的多元碰撞，探讨人类繁荣前景的替代性方案。

Circular Design:
Wine Box Reuse Program

Partners: College of Design and Innovation, Godolphin, Wanhua Chemical

Support: Ellen MacArthur Foundation

Academic Hosts: Hao Yang, Jiao Mo

About: To a large extent, our economy thrives on fast moving consumer goods. Can these items, produced for a short use, have a longer life cycle and continuous create value before the nutrients they contain are returned to the natural system? The wine boxes archipelago and the plastic bamboo forest create the theme pavilion space. How can they be reused, remanufactured and recycled after the exhibition? The workshop will explore alternative solutions to the future of human prosperity through the collision of circular economy, technological innovation, human oriented insight and creative imagination.

声音雕塑——机械臂工作坊

合作伙伴：声音实验室

学术主持：周洪涛，郑康奕

活动简介：本工作坊通过数字算法，将声音转换成形态。并通过6轴工业机械臂三维切割，现场完成雕塑作品。参与者可以尝试用不同的声波控制几何图形，理解声音可视化原理，并创作具有个性化标签的艺术作品。作品以高密度泡沫作为材料，工业机械臂结合数字化设计与生产，高效、低成本和即时地完成了参与者的创意。

Sculpture of Sound — Robotic Arm Workshop

Partners: Sound Lab

Academic Hosts: Hongtao Zhou, Kangyi Zheng

About: This workshop convert sound into form by algorithms. Use the 6-axis industrial robotic arm cutting it by the three-dimensional, and the sculpture works are completed on site. Participants can experiment with different sound waves to control geometric shapes, understand the principles of sound visualization, and create artwork with personalized labels. The works are made of high-density foam, and the industrial robot arm combines digital design-production to complete the participants' ideas efficiently, cost-effectively and instantly.

食物设计工作坊   Food Design Workshop

合作伙伴: 好公社, Punchline

学术主持: 陆洲, 魏佛兰

活动简介: 邀请擅长食物保存和发酵的达人，从各个角度来探讨下食物保存技术、发酵文化以及食物背后的故事。发酵作为一种千百年流传下来的食物保存经验，既可以很好地解决食物浪费问题，又让时间给食物添上多样别致的风味。同时，发酵产生的活动着的益生菌也对人身体非常有益，在今日，更是一种很环保、很可持续的处理和保存食物的行为。

Partners: NICE COMMUNE, Punchline

Hosts: Zhou Lu, Francesca Valsecchi

About: We invited experts in food preservation and fermentation to discuss food preservation techniques, the culture of fermentation and the stories behind. Fermentation as an old food preservation practice that not only solves the problem of food waste, but also adds a variety of flavours to food over time. Meanwhile it is a very environmentally friendly and sustainable way of handling and preserving food.

花开中国：从1到1 000到1 0000

合作伙伴：同济大学社区花园与社区营造实验中心，四叶草，全国SEEDING联盟，召集社区花园团园行动的团长们，第二届全国社区花园社区参与行动的队员们

学术主持：刘悦来，刘畅

活动简介：社区花园从上海开始的第一个火车菜园，到南宁的1 000个老友花园，到全国联盟的发展历程及最新计划——万人参与的团园行动SEEDINGARDENS和第二届全国社区花园设计营造竞赛与社区参与行动，现场将发布团园行动超过1 000份问卷的结果，并对于竞赛阶段性进展进行现场发布，针对团园行动中出现的疑难问题进行现场讨论与互动，以支持人民参与的公共空间生产行动。

Community Garden Blooming China: From 1 to 1,000 to 10,000

Partners: Lab of Community Garden and Community Empowerment, Tongji University; Shanghai Clover Nature School; National Alliance of SEEDING; all leaders of community garden program, and all members of 2nd nationwide community garden program

Academic Hosts: Yuelai Liu, Chang Liu

About: The development history of community garden from the first train garden in Shanghai to 1,000 old-friend garden in Nanning to the National Alliance, and the new plan, the 10,000 participants SEEDINGARDENS action and the Second National Community Garden Design and Construction Competition and Community Participation Action. We will release the results of over 1,000 questionnaires and report the stage process of the competition on-site. Moreover, we will also share and discuss about the experience in the garden community process, which as a representative case to support the public space production with people's participation.

健康设计 / Design for Health

合作伙伴：同济大学设计创意学院，同济大学医学院，同济大学中意学院，上海国际设计创新研究院，同济黄浦设计创意中学，上海交通大学医学院附属上海儿童医学中心，上海音乐学院音乐工程系，NICE2035未来生活原型街，NICE公社，CHEER Design实验室，鸟鸣电台

学术主持：刘震元，刘毅

活动简介：将健康作为设计的对象，来自设计、医疗、艺术、管理、工程、教育、社会学、心理学等领域跨学科的专家小组，将围绕"健康状态的策育""健康知识的传达""健康技术的体验"和"健康决策的制定"等复杂社会技术议题，大开脑洞，协同共创，在相互对话和激发过程中，完成一次对有关健康设计的高强度探索。

Partners: College of Design and Innovation, Tongji University; Tongji University School of Medicine; Sino-Italian Campus, Tongji University; Shanghai International Institute of Design and Innovation, Tongji University; Tongji-Huangpu School of Design and Innovation; Shanghai Children's Medical Center, Shanghai Jiaotong University School of Medicine; Department of Music Engineering, Shanghai Conservatory of Music; NICE2035 Living Line; NICE Commune; CHEER Design Lab, Birds Chorus Radio Station

Academic Hosts: Zhenyuan Liu, Yi Liu

About: Taking health as the object of design, a group of interdisciplinary experts from design, medicine, art, management, engineering, education, sociology and psychology will dive into complex socio-technical issues about human health, covering "Health Curation","Heath Communication","Health Technology Experience" and "Health Decision-making". By open-mind conversation and co-creation, the group is going to make an intensive exploration in terms of designing for health.

Fablab儿童友好工作坊:"跟着教授玩AI"

Kids AID

合作伙伴:同济大学设计创意学院,尚想实验室,Fablab O Shanghai数制工坊,艾厂,同济黄浦设计创意中学

学术主持:康思大,丁峻峰,陶斌,陈得恩

活动简介:"跟着教授玩AI"是一个面向8—15岁的孩子学习关于人工智能和设计与艺术的项目。同济大学设计创意学院尚想实验室联合Fablab O Shanghai数制工坊一起,为城市的孩子们开发"线下"工作坊,为远郊和农村的孩子们提供"线上"课程,并为孩子最终作品提供展览、活动及比赛平台。和常规编程课程不同,本工作坊的独特之处在于:我们用事例、故事、神话、谜题、游戏、迷因模仿或任何能触发他们的思维方式的内容,来启发孩子们人工智能的逻辑,驱动他们人工智能的想象力和思考。此次工作坊的重点是关注自我,以及人工智能如何帮助我们表达情感。

Partners: College of Design and Innovation at Tongji Univerisity, Shang Xiang Lab, Fablab O Shanghai, AAI, Tongji-Huangpu School of Design and Innovation

Academic Hosts: Kostas Terzidis, Jeff Ding, Bin Tao, De'en Chen

About: Kids AID is a about teaching AI using design and art for kids between 8 and 15 years old. Together with Fablab O Shanghai we do offline workshops for urban kids, online classes for unprivileged kids, exhibitions, and competitions. There are many courses on AI coding, but the uniqueness of our education program is that we teach the logic of AI using examples, stories, myths, puzzles, games, memes, or anything that would trigger their mind to think differently. To think like Artificial imagination. This workshop focuses on the self and how AI can help us understand who we are. The outcome will be a self-portrait.

致谢
Acknowledgements

中国工业设计协会	China Industrial Design Association
国务院第八届设计学学科评议组	The 8th Design Discipline Evaluation Group of the State Council
WDO世界设计组织	World Design Organization
ICoD国际设计组织联合会	International Council of Design
CUMULUS国际艺术、设计与媒体院校联盟	CUMULUS Association
DESIS社会创新与可持续设计联盟	DESIS Network
瑞安新天地	Shui On XINTIANDI
尤伦斯当代艺术中心	UCCA Center for Contemporary Art
设计上海	Design Shanghai
设计互联	Design Society
博埃里建筑设计事务所	Stefano Boeri Architetti
米兰三年展美术馆	Triennale di Milano
上海市工业设计协会	Shanghai Industrial Design Association
上海市青年创意人才协会	Shanghai Young Creative Talents Association
上海市青少年创意设计院	Shanghai Youth Institute of Design and Innovation
上海市工业设计研究院	Shanghai Industrial Design and Research Institute
同济大学设计创意学院	College of Design and Innovation, Tongji University
东华大学设计学院	College of Fashion and Design, Donghua University
上海大学美术学院	Shanghai Academy of Fine Arts, Shanghai University
华东理工大学艺术设计与传媒学院	School of Art, Design and Media, East China University of Science and Technology
上海工程技术大学艺术设计学院	School of Art and Design, Shanghai University of Engineering Science
上海应用技术大学艺术与设计学院	School of Art and Design, Shanghai Institute of Technology
上海创新创意设计研究院	Design Innovation Institute Shanghai
中国设计智造大奖	Design Intelligence Award
中国设计大展	China Design Exhibition
"好设计"奖	Good Design Award
戴森设计大奖	The James Dyson Award
艾伦·麦克阿瑟基金会	Ellen MacArthur Foundation
同济大学建筑设计研究院（集团）有限公司	Tongji Architectural Design(Group) Co., Ltd.
上海同济城市规划设计研究院	Shanghai Tongji Urban Planning & Design Institute Co., Ltd.
中国建筑第八工程局有限公司	China Construction Eighth Engineering Division Co., Ltd.
特赞（上海）信息科技有限公司	Tezign
马努艺术	Manu.Art
阿旺特	AVARTE
设计丰收	Design Harvest
好公社	NICE COMMUNE
小红书	Xiaohongshu
《设计，经济与创新学报》	*She Ji, the Journal of Design, Economics and Innovation*
R.I.S.E.可持续时尚创新平台	R.I.S.E. Sustainable Fashion Innovation Platform

Acknowledgments

第二部
Part Two

2023年"设计无界,造化万象"世界设计之都大会主题展——
The 2023 "Design Beyond Creativity" World Design Cities Conference Theme Exhibition—

创意社群:嵌入式、可感知与高交互
NICE Commune: Embedded, Sensible and Interactive

策展团队
Curatorial Team

总策展人
Chief Curator

娄永琪
Yongqi Lou

教授 同济大学副校长
英国皇家艺术学院荣誉博士
瑞典皇家工程科学院院士
Professor and Vice President of Tongji University
Honorary Doctor of Royal College of Art
Fellow of Royal Swedish Academy of Engineering Sciences

娄永琪教授，博士，同济大学副校长，全国设计研究生教育指导委员会主任委员；长期致力于社会创新和可持续设计实践、教育和研究；在同济大学设计创意学院自创立（2009年）至2022年其卸任院长期间，推动学院成为世界一流设计学院；先后担任CUMULUS国际艺术设计院校联盟副主席、WDO世界设计组织执委、维也纳应用艺术大学国际咨询委员会主席。他是《设计、经济与创新学报》（爱思唯尔出版）的创始执行主编，也是《设计问题》（麻省理工学院出版）的编委，还是2021年国际设计研究大会联合主席，并受邀在2018年香港设计营商周、2017年国际室内建筑师/设计师团体联盟代表大会、2016年美国工业设计协会国际大会和2015年国际人机交互大会等著名会议担任主旨演讲人。他的作品和研究成果在芬兰赫尔辛基设计博物馆、意大利米兰三年展博物馆等处展出；他于2019年当选瑞典皇家工程科学院院士，于2023年获英国皇家艺术学院荣誉博士学位。

Prof./Dr. Yongqi Lou is a vice president of Tongji University and the Chair of the China National Graduate Education Steering Committee for Design. Lou has been at the forefront of the education, research, and practice of design for social innovation and sustainability. He has played a crucial role in the success of the College of Design and Innovation (D&I) of Tongji University, since its inception in 2009 and through his deanship till 2022. Lou has served on various boards, including CUMLUS, ICSID/WDO, DESIS, and Angewandte (Vienna). He is the founding executive editor of *She Ji: The Journal of Design, Innovation, and Economics* (published by Elsevier) and serves as an editorial board member of *Design Issues* (MIT Press). He was the co-chair of IASDR 2021 and was invited as a keynote speaker at prestigious events such as BODW 2018, IFI 2017, IDSA 2016, and ACM SIGCHI 2015. His works have been exhibited in the Design Museum of Helsinki, the Triennale Design Museum in Milan, and other renowned institutions. Lou was elected as a fellow of the Royal Swedish Academy of Engineering Sciences (IVA) in 2019, and in 2023, he received an honorary doctorate from the Royal College of Art.

策展人 / Curators

按首字拼音排序 / Sorted by first letter of Chinese pinyin

安东·西比克
Aldo Cibic

范凌
Ling Fan

康思大
Terzidis Constantinos

刘洋
Yang Liu

刘翠
Zhao Liu

刘震元
Zhenyuan Liu

陆洲
Zhou Lu

莫娇
Jiao Mo

萨维里奥·西利
Saverio Silli

苏雅默
Jarmo Suominen

魏佛兰
Francesca Valsecchi

赵华森
Huasen Zhao

赵世笃
Shijian Zhao

张周捷
Zhoujie Zhang

郑康奕
Kangyi Zheng

周洪涛
Hongtao Zhou

助理策展人   Assistant Curators

黄粒莹
Liying Huang

纪丹雯
Danwen Ji

刘畅
Chang Liu

王依琳
Yilin Wang

空间及平面设计   Spatial and Graphic Design

丁卓媛
Zhuoyuan Ding

杜钦
Qin Du

郭泠
Ling Guo

尤优
You You

策展序言
Curatorial Statement

2023年WDCC主题展是本次大会主题"设计无界，造化万象"的学术演绎。主题中的"造"是指"制造"和"创造"，是人类改造世界最主要的方式；"化"是"变革""融入""创未来"的意思；而"万象"既是指设计本身的气象万千，更是指设计全方位赋能的作用和价值。主题中的"造化"是创造的最高境界，它所表意的"自然"，是人类智能和人工智能（如自然语言模型）最高等级的创造逻辑。

作为有全球影响力的世界设计之都，上海不应该仅仅满足于出色地回答问题，而应该更加主动地提出新问题。通过思想、作品、品牌、产业、生态，逐步凝结一个不断壮大、主动担当、引领未来的创意社群。上海举办世界设计之都大会，就是要催生和支持这个社群的成长，并为他们提供一个思想交流、创意互鉴、未来洞察的全球共创共享平台。

如果说2022年的WDCC主题展重在提出问题、凝结社群，那么2023年主题展，则更侧重于呈现这个创意社群的万千气象，包括他们的思考、成果以及社群的互动。"城市是最好的大学，上海这座城市就是最好的设计大学！"展览以NICE2035未来生活原型街区这一同济大学设计创意学院与杨浦区四平路街道"区校合作、三区联动"的共创成果为蓝本，用"盒子社区"的方式展出嵌入NICE2035社区中的同济-阿斯顿·马丁创意实验室、声音实验室、机械臂实验室、AIGC实验室、材料实验室、首饰实验室、微装配实验室、食物实验室、城市科学实验室、安东·西比克工作室等十余个空间，以及与之相关的设计人物、设计作品和设计事件。八个与实验室紧密相关的体验式工作坊将成为"活态展品"，贯穿整个展期，与参观者充分互动。整个展览场景呈现了打破大学和城市边界后，自发生长出来的国际化创意社群的勃勃生机，也彰显了上海"城市处处有设计""寓设计于生活"的理念、主张和行动。

The theme exhibition of WDCC2023 is an academic interpretation of the conference theme. The theme in Chinese, " 设计无界，造化万象 (She Ji Wu Jie, Zao Hua Wan Xiang)" embodies the concept of "Design Beyond Creativity". In Chinese cultures, " 造 (Zao)" refers to both "manufacturing" and "creation", representing the primary ways in which humans transform the world." 化 (Hua)" signifies "transformation", "integration"and "generation", while " 万象 (Wan Xiang)" represents the diverse aspects of design itself and, more importantly, the comprehensive empowerment and value it provides. The term " 造化 (Zao Hua)" in the theme represents the highest realm of creation, while " 自然 (Zi Ran)" signifies the highest level of creative logic for human intelligence and artificial intelligence (such as natural language models).

As a globally influential World Design Capital, Shanghai should not only excel at answering questions in a right way, more important, Shanghai should actively raise or reframe visionary questions. Through creating new ideas, new works, new brands, new industries, and new ecosystems, it gradually consolidates a creative community that can proactively lead the future. The Shanghai World Design Cities Conference aims to nurture and support the growth of this community and provide them with a global collaborative and sharing platform for exchanging experience and inspiring futures insights.

If the emphasis of the WDCC theme exhibition in 2022 was on posing questions and fostering community cohesion, the 2023 theme exhibition is more focused on presenting the myriad manifestations of this creative community, including their thoughts, achievements, and interactions. "The city is the best university, and Shanghai is the best design university!" The theme exhibition takes the case of NICE2035 Living Line as the blueprint, which is a collaborative initiative between the College of Design and Innovation at Tongji University and the Siping Road Subdistrict in Yangpu District. It showcases more than ten laboratories, including the Tongji-Aston Martin Creative Lab, sound lab, robotic arm lab, AIGC lab, material lab, jewelry lab, micro assembly lab, food lab, city science lab, and Aldo Cibic Studio, embedded in the NICE2035 community in a "box community" format. It also presents related designers, design works, and design events. Eight immersive workshops closely related to the laboratories will serve as "live exhibits" throughout the exhibition period, allowing for full interaction with visitors. The exhibition presents a scenario reflexing the vibrant vitality of an international creative community that has spontaneously thrived via breaking the boundaries between the university and the city. It also reflects Shanghai's design culture and attitude of "everywhere design in the city" and " imbedded design into daily life".

策展结构：嵌入社区的创新实验室集群
Curatorial Structure: Innovation Labs Embedded in Communities

今年的主题展以"盒子社区"的形式，通过"小中见大"的方式模拟位于NICE2035未来生活原型街中的诸多实验室。每个盒子都代表了一个实验室。在盒子的内部和周边，陈列的是实验室的创新产出和创新氛围。从空间上来看，展厅里形成了一条可供徜徉的街道。由此，参观者也成为了展品之外展览的一个活态部分。

本图册接下来的内容，是以一个个实验室为结构主线，呈现每个实验室在近年来产出的代表性作品和创新思考。同时本图册也收录了展览期间，每个实验室举办的协同创新工作坊。

This year's theme exhibition is in the form of a "box community", simulating the laboratories located in the NICE2035 community through a "small to big" approach. Each box represents a laboratory. Inside and around the box, the innovative output and atmosphere of the laboratory are displayed. In terms of space, a street has been formed in the exhibition hall where visitors can wander around. In addition to the exhibits, visitors have also become a lively part of the exhibition.

The following content of this brochure is structured around each laboratory, presenting the representative works and innovative ideas produced by each laboratory in recent years. It also includes collaborative innovation workshops held by each laboratory during the exhibition period.

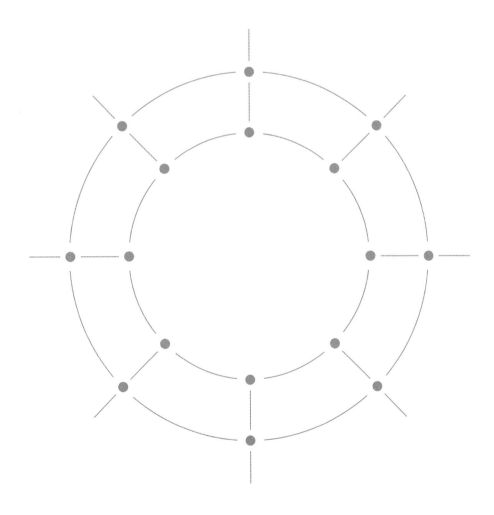

实验室介绍
Labs' Introduction

盒子社区
Box Community

盒子社区是由青山周平和杜晓峰发起的一种新型社区形式，通过以盒子为主要载体的模块化解决策略，为社区成员提供多样化的服务和活动，促进社区成员之间的交流和互动，同时为社区居民提供便利的生活服务。2023 年 WDCC 主题展用盒子社区的形式，模拟了同济大学与四平路街道共建的 NICE2035 未来生活原型街，小中见大地呈现上海城市"处处有设计"的鲜活景象：拓展到社区的实验室和社区生活紧密相连，每个盒子都代表一个实验室，盒子之间，以及社区与整个展场之间，一直在发生互动和意想不到的故事。上海这座城市才是最好的创新孵化器和最好的设计大学！

The Box Community is a new type of community concept initiated by Shouhei Aoyama and Xiaofeng Du. Through a modular solution strategy centered around "boxes", it provides a variety of services and activities for community members, fostering communication and interaction among them while also offering convenient living services for residents. The 2023 WDCC theme exhibition used the Box Community model to simulate the NICE2035 Living Line, a joint initiative between Tongji University and Siping Road Subdistrict. This concept subtly yet vividly showcased the lively scenes of "design everywhere" in Shanghai's urban fabric: laboratories extend into the community and are closely linked with daily life. Each box represents a laboratory, and the interactions—both among the boxes and between the community and the entire exhibition space—are constantly unfolding, leading to unexpected and fascinating stories. Shanghai itself is the best innovation incubator and the best design university!

同济-阿斯顿·马丁创意实验室
Tongji-Aston Martin Creative Lab

同济-阿斯顿·马丁创意实验室是英国超豪华汽车制造商阿斯顿·马丁首个设置在英国本土之外的设计工作室。该创新中心主要对创意设计的前沿进行探索，研究全球千禧一代的需求以及未来设计的发展趋势。

Tongji-Aston Martin Creative Lab is the British brand's first design studio outside of the UK. It focuses on not only the automobile field but also art design, future community, traffic, luxury lifestyle, and interior decoration to offer a glimpse into future life.

声音实验室
Sound Lab

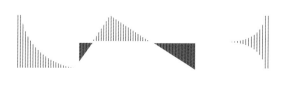

声音实验室由娄永琪教授与跨界音乐家朱哲琴联合发起。以声音为原动引擎，结合算法设计、媒体交互等技术，以国际化、前瞻性和实验性为特色，尝试各种跨界、跨学科、跨文化的前沿研究与社会公共应用实践。

Sound Lab is jointly launched by Prof. Yongqi Lou and cross-border musician Zheqin Zhu. Using sound as the driving engine, combined with algorithm design, media interaction and other technologies, we attempt to explore various cross-border, interdisciplinary, and cross-cultural cutting-edge research and social public application practices.

Fablab O Shanghai 数制工坊
Fablab O Shanghai

Fablab O Shanghai
数制 工坊

Fablab O Shanghai是同济大学设计创意学院主办的小型工作坊，通过提供数字制造工具，使用户能够"制造（几乎）任何东西"。它是麻省理工学院领导的Fablab全球网络的重要成员，旨在为用户提供那些通常在大规模工业化生产当中应用的技术，让用户能够体验"个人制造"。

Fablab O Shanghai is a small-scale workshop hosted by the College of Design and Innovation at Tongji University, offering digital fabrication tools and enabling users to "make (almost) anything". It is part of an international network led by MIT whose aim is to give access to technologies normally found in mass-production factories, allowing users to experience "personal manufacturing".

安东·西比克工作室
Aldo Cibic Studio

ALDO CIBIC

1955年，安东·西比克出生于意大利维琴察。1981年，作为索特萨斯事务所的合伙人之一，他参与了孟菲斯流派的创立。2018年起，他成为同济大学荣誉教授。2023年，他被授予意大利共和国功勋骑士勋章。

Aldo Cibic, was born in Vicenza, Italy, in 1955. In 1981, as a partner in Sottsass Associati, he was a founding member of Memphis. Since 2018, Aldo Cibic has been an honorary professor at the Tongji University, Shanghai. In 2023 he was awarded the honor of Knight of the Order of Merit of the Italian Republic.

祎设计
Yi Design

祎设计将工业陶瓷废料转化为可用于装饰、结构,且可定制的各类瓷砖、建筑砖,为室内和建筑行业提供可循环选项。

Yi Design offers a circular solution for the interior and architecture industry, by transforming industrial ceramic waste into decorative, structural, and customizable tiles and bricks.

同济-麻省理工上海城市科学实验室
Tongji-MIT City Science Lab @Shanghai

本实验室由同济大学上海国际设计创新学院与麻省理工学院媒体实验室共同建立,聚焦未来可持续生活方式和产业转型,针对城市交通、智慧生活、社区营造等重点领域,以数据和前沿技术支持设计、科技和产业的创新与转型。

Tongji-MIT City Science Lab @Shanghai is jointly established by Tongji University Shanghai International College of Design and Innovation and MIT Media Lab. We focus on future sustainable lifestyles and industrial reform in key areas including urban transportation, smart life, and community building. We use data and leading technologies to support innovation and transformation of design, technology, and industries.

SustainX 可持续未来设计研究中心
SustainX Lab

SustainX
可持续未来设计研究中心

SustainX可持续未来设计研究中心由娄永琪教授领导，面向可持续设计创新，聚焦《DesignX议程》中提出的"复杂社会技术系统设计"问题，系统地思考在人工智能时代环境、健康、交通、人居、产业、教育等问题的解决中，设计的新角色、新方法、新路径和新工具。

The SustainX Lab, led by Prof. Yongqi Lou, is dedicated to sustainable design innovation. It focuses on researching the problems of "complex sociotechnical systems design" proposed by the "DesignX Agenda". The center systematically explores the new roles, new methods, new pathways, and new tools in the design of solutions for various issues such as environment, health, transportation, living spaces, industry, and education in the era of artificial intelligence.

好公社
NICE COMMUNE

NICE COMMUNE 好公社
DESIGN Harvests 设计丰收

好公社（NICE COMMUNE）是NICE2035未来生活原型街的社区引擎，是一个集共享厨房、精品咖啡、展示长廊和创意办公为一体的多变空间。通过将外部社群的新事物与社区"原住民"的生活和文化相结合，打造一个融合各类社群的文化交流载体、一方无限可能的孵化地、一处未来生活的原型场、一批扎根社群的"生活实验室"、一处大学知识溢出的基地。好公社倡导社区作为创新的源头，尊重多元化、包容性和共创共享。

NICE COMMUNE is the community engine of the NICE2035 Living Line, offering a versatile space that combines a shared kitchen, boutique coffee shop, exhibition corridor, and creative offices. By integrating new ideas from external communities with the lives and culture of local residents, it aims to create a platform for cultural exchange among diverse groups, an incubator of endless possibilities, a prototype space for future living, a series of "living labs" deeply rooted in the community, and a hub for the dissemination of university knowledge. NICE Commune advocates for the community as a source of innovation, embracing diversity, inclusivity, and co-creation.

同济DESIS实验室
Tongji DESIS Lab

同济大学社会创新与可持续设计实验室是全球社会创新与可持续设计联盟（DESIS NETWORK）的重要成员之一。作为立足于同济大学设计创意学院的具有国际与前瞻视野的科研平台，实验室长期聚焦于可持续生活方式的研究，通过"主动设计"积极介入城乡互动、社区营造、创新教育等社会领域，实现设计驱动社会创新与改变。

The Tongji DESIS Lab is a key member of the global DESIS Network. As an internationally oriented research platform based at the College of Design and Innovation at Tongji University, the lab focuses on sustainable lifestyle research. Through "proactive design", it actively engages in social domains such as urban-rural interaction, community building, and innovative education, aiming to drive social innovation and change through design.

JALAB首饰实验室
JALAB

JALAB首饰实验室创立于2013年，致力于推动新兴技术与传统产业的交流与整合。实验室课程涵盖时尚设计、数字工艺、智能可穿戴等主题。自成立以来，已经同众多知名时尚品牌开展合作，并在丰富的实践中不断提升设计的品质与工艺流程。此次盛会，实验室带来的是其与路威酩轩香水化妆品中国/迪奥（LVMH Beauty China/DIOR）共创的可持续时尚课题研究成果，同时由瑞安低碳办公联盟共同策展。

JALAB Lab was founded in 2013 to promote the exchange and integration of emerging technologies and traditional industries. Lab courses cover topics such as fashion design, digital craftsmanship, and smart wearables. Since its establishment, it has cooperated with many well-known fashion brands, and continuously improved the quality and process of design in rich practice. The event Lab brought the results of its research on sustainable fashion co-created with LVMH Beauty China/DIOR and Xintiandi Sustainable Solutions.

同济大学时尚与设计创新中心
FINE Center

同济大学时尚与设计创新中心是由国家级海外引进学者牵头的高层次文教产研相结合的科研中心，以"大时尚"为语境，设计与艺术为驱动力，专注当下和未来品质生活方式等相关主题的研究、实践与策划管理，以"创新领袖"为导向的多功能国际国内协作发展平台。

FINE Center is a scientific research center combines high-level culture, education, production and research led by national-level oversea scholars. It is a leading multi-functional international and domestic collaborative development platform with the context of "big fashion" and the driving force of design and art, focusing on research, practice, and planning management of current and future quality lifestyles and other related topics, guided by "Creative Leadership".

材料与应用创新实验室
Material and Application Design Laboratory

材料与应用创新实验室关注材料创新和与之相匹配的加工应用技术，连接新型数字加工与传统手工艺，从而拓展创新的范畴。实验室秉承可持续原则，通过探索材料再创的无限可能，源源不断地为产品及其服务提供有意义的价值。

Material and Application Design Laboratory focuses on innovative material and its matching application processing technology, by associating computer numeric manufacture with traditional crafts to enhance the spectrum of designer creativity. Adhering to the principle of sustainability, by exploring the infinite possibilities of material regeneration, the Lab will continuously explore meaningful value for products and services.

造物实验室
Making Lab

同济造物实验室由周洪涛教授引领，旨在追求研究导向的艺术和设计创造，并将突破边界的创作成果无缝融入真实世界的挑战之中。造物实验室肩负着一个强大的教育使命，就是让年轻学生获得尖端的造物知识与技能，将想象力转换成现实。

The Tongji Making Lab, led by Prof. Hongtao Zhou, pursues the vision of research-guided design and making of artistic and designed objects to seamlessly expend their boundaries into real world challenges. It has a strong education mission to enable young students to gain cutting-edge knowledges and skills of tangible making to empower their imaginations into realities.

设计人工智能实验室
Design A.I. Lab

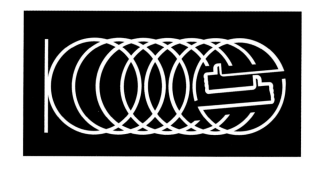

同济大学设计人工智能实验室致力于生成式人工智能与设计学的交叉学科应用研究。实验室与科技企业和社会团体合作，实现产、学、研、创的转化；通过人工智能技术实践社会创新，如人工智能的少儿科普、人工智能的传统手工艺保护等。

Tongji Design A.I. Lab is committed to the interdisciplinary application research of generative artificial intelligence and design. The lab cooperates with technology companies and social groups to integrate learning, research and innovation. The lab also conducted social innovation projects using artificial intelligence technology, such as AI for children, AI for traditional cultural heritage, etc.

尚想实验室
Shang Xiang Lab

尚想实验室专注于基于人工智能和参数化的设计——使用不同元素的智能组合，实现复杂形式的合成。除了学术研究，实验室还致力于儿童人工智能的教育和知识分享。

The Shang Xiang Lab focuses on artificial intelligence and parametric design using exhaustive combinations of elements in the synthesis of complex forms. In addition to academic research, the laboratory is also dedicated to the education and knowledge sharing of Kid's AI.

艺术与人工智能实验室
Art & Artificial Intelligence Lab

同济大学设计创意学院艺术与人工智能实验室于2021年成立，简称AAI。实验室致力于艺术和人工智能技术的交叉学科应用研究，作为一个探索平台，实验室汇聚人工智能艺术领域的艺术家、科学家和技术专家等，共同研究智能算法在创意和艺术领域的应用，并最终实现产、学、研、创的转变。

The Art & Artificial Intelligence Lab (AAI) at the College of Design and Innovation, Tongji University, was established in 2021. The lab is dedicated to interdisciplinary research at the intersection of art and AI technology. As an exploratory platform, it brings together artists, scientists, and technical experts in the field of AI art to collaboratively study the application of intelligent algorithms in creative and artistic domains, ultimately aiming to bridge the gap between industry, academia, research, and innovation.

载运工具与系统创新设计实验室
Next Mobility Lab

载运工具与系统创新设计实验室隶属于同济大学设计创意学院，战略性地结合了同济大学在设计和国际化方面的优势。实验室以"智能体验"和"系统可持续性"为核心，致力于应对国家、社会和经济在未来交通与出行领域的新兴创新设计需求。

The "Next Mobility Lab", part of College of Design and Innovation of Tongji University, strategically combines the university's strengths in design and internationalization. With a core focus on "intelligent experiences" and "system sustainability", it addresses emerging innovative design needs in future transportation and mobility for the nation, society, and the economy.

机械臂实验室
Robotic Arm Lab

位于NICE2035未来生活原型街的机械臂实验室由同济大学设计创意学院与Kuka机械臂共同建立。现场展出的机械臂与AI协助绘画，基于人工智能算法，学习与模拟人类情感，采用计算机视觉、非真实渲染等技术，完成计算与自动控制。通过模拟学习人的行为，六轴机器人进行实时颜色组合调整，从而完成整个作画过程。现场与观众的多重交互，鼓励人们思考：机器是帮助人类从重复、烦琐的劳动中解脱出来？还是反过来制约人类行为和思想？

Located on NICE2035 Living Line, the Robotic Arm Lab is a collaborative venture between College of Design and Innovation of Tongji University and Kuka Robotics. The lab showcases a robotic arm working with AI-assisted painting. Utilizing artificial intelligence algorithms, the system learns and simulates human emotions through computer vision and non-realistic rendering technologies, achieving precise computation and automated control. By simulating human behavior, the six-axis robot adjusts color combinations in real time to complete the painting process. The interactive experience with the audience prompts reflection on whether machines serve to free humans from repetitive and tedious tasks or conversely, constrain human behavior and thought.

实验室参展作品
Introduction of Labs' Works

印象·马丁光绘艺术表演摄影
Impression of AM Light Art Performance Photography

居民和学生 | 同济－阿斯顿·马丁创意实验室 | 2020
Residents and Students | Tongji-Aston Martin Creative Lab | 2020

与社区居民共创，用光绘为阿斯顿·马丁的首款SUV——DBX勾勒出艺术的轮廓，营造炫酷的视觉体验。

Co-created with community residents, light painting was used to outline the artistic contours of Aston Martin's first SUV, the DBX, creating a stunning visual experience.

越过光的形状
Shape of Light

闻一然 | 同济 - 阿斯顿·马丁创意实验室 | 2020
Yiran Wen | Tongji-Aston Martin Creative Lab | 2020

"越过光的形状"是以阿斯顿·马丁的车标为基础,从平面图形衍生至三维空间的数字雕塑。雕塑将实体与光影结合,在点光源的照射下可在平面上投射出阿斯顿·马丁车标的形状。造型灵感源于阿斯顿·马丁GT跑车飞驰于蜿蜒的赛道:从原点出发,然后经过长途跋涉与极限弯道后回到原点,充满速度感与运动感。该作品以虚实结合的手法体现阿斯顿·马丁"惟·美"的品牌核心。

"Shapes of Light" is a digital sculpture based on the Aston Martin logo, extending from a two-dimensional graphic to a three-dimensional space. The sculpture merges the tangible with light and shadow, allowing a point light source to project the shape of the Aston Martin logo onto a flat surface. The design is inspired by the Aston Martin GT sports car speeding along a winding track: starting from the origin, traversing long journeys and extreme curves, and finally returning to the starting point, embodying a strong sense of speed and motion. This work reflects Aston Martin's brand values through the interplay of reality and illusion.

智绣，绣出未来座舱：探索非遗与智能织物的创新应用

Smart Textiles Embroidering the Future Cockpit: Exploring the Integration of Smart Textiles in Next Mobility

同济－阿斯顿·马丁创意实验室 | 2021
Tongji-Aston Martin Creative Lab | 2021

本作品以智能座舱为载体，探索将智能织物与非遗刺绣织布等中国非物质文化遗产相融合，并将之应用于未来出行场景中。本设计希望建立起上海刺绣标布非遗传承人及其机构、同济大学设计创意学院与阿斯顿·马丁三方共创的机制，促进上海非遗刺绣与智能科技相结合，探索出新的应用场景。

This work explores to use the intelligent cockpit as a platform to integrate smart fabrics with Chinese intangible cultural heritage, such as embroidery and weaving, applying them to future mobility scenarios. The design aims to establish a co-creation mechanism among Shanghai's embroidery heritage inheritors and their institutions, the College of Design and Innovation at Tongji University, and Aston Martin. This collaboration seeks to promote the fusion of Shanghai's traditional embroidery with smart technology, paving the way for new application scenarios.

高山流水
Mountain Stream

郁新安 | 声音实验室 | 2018
Xin'an Yu | Sound Lab | 2018

本作品使用了著名琵琶曲《高山流水》，描绘了"峨峨兮若泰山"和"洋洋兮若江河"。作品运用算法将琴声转化为溪水的纹理，主要探索这一主题下的声音与图像的关联，并且使用数字技术，将前者凝固为后者。与不断重复低频的现代电子乐不同，传统国乐比如琵琶曲的频段分布广泛而跳跃，因此不同的曲子所对应的可视化的逻辑也不同。本作品针对传统国乐特殊的频段分布，对乐曲进行分段采样，并用采集的数据驱动漩涡状纹理，获得类似水墨或木器表面，却不断变换着的纹理。

This work uses the renowned pipa piece *Mountain Stream* to convey the grandeur of "Lofty as Mount Tai" and the vastness of "Expansive as a River". An algorithm transforms the pipa's sound into flowing water textures, exploring the relationship between sound and imagery. Digital technology solidifies sound into visual form. Unlike modern electronic music's repetitive low-frequency beats, traditional Chinese music, like pipa compositions, has a broad and dynamic frequency distribution. Thus, visualization should vary with each piece. By segmenting the music based on its frequency, the sampled data drives swirling textures, creating visuals resembling shifting ink or wood grain, constantly evolving.

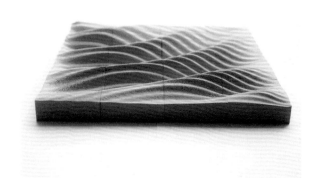

数字之都
Digital City

张屹南，郁新安，谭丞超 | 声音实验室 | 2021
Yinan Zhang, Xin'an Yu, Chengchao Tan | Sound Lab | 2021

由海关钟声生成的数字化虚拟都市在美学上融合了实体上海市特征与算法设计的风格，表现出2021世界人工智能大会"科艺同频"的审美追求。

The digital virtual city, generated from the sound of the Customs House bell, aesthetically merges the characteristics of physical Shanghai with algorithmic design styles, embodying the aesthetic pursuit of "technology and art in harmony" showcased at the 2021 World Artificial Intelligence Conference.

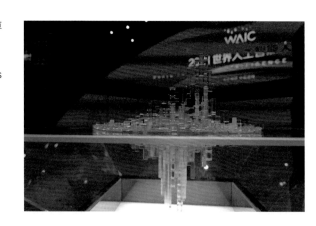

涟漪
Ripple

郑康奕 | 声音实验室 | 2019
Kangyi Zheng | Sound Lab | 2019

本作品探索了声音可视化，并以造物的方式呈现。本作品基于阿斯顿·马丁的品牌特性，将DB11引擎启动的声音，通过数字成像技术及计算机算法演化转换到三维空间。本设计以仪表盘作为基座，从圆心出发的波纹起伏则对应引擎声波的频率高低。随着时间推移，声音渐渐消失，圆盘外围越来越平缓。纵向对应声波频率，横向对应时间维度。

This work explores the visualization of sound through tangible creation. Based on the characteristics of the Aston Martin brand, the sound of the DB11 engine starting is transformed into a three-dimensional space using digital imaging technology and computer algorithms. The design uses the dashboard as a base, with ripples emanating from the center corresponding to the varying frequencies of the engine's sound waves. As the sound gradually fades, the outer edges of the disc become increasingly smooth. The vertical dimension corresponds to the sound wave frequency, while the horizontal dimension represents the passage of time.

Percuino：用非人类感官唤醒感知觉
Percuino: Awakening Perception Through Non-Human Senses

高培中 | Fablab O Shanghai 数制工坊 | 2022
Peizhong Gao | Fablab O Shanghai | 2022

在这个信息传递效率至上的时代中，视觉几乎构建了数字时代的全部现实。本作品的设计师希望人们重新拾起丰富的感知维度，更多地关注电子屏幕之外的城市自然，追求感知的丰富度。通过与动物交换感官，使用者尝试用非人类感官感受熟悉的城市自然。三块可替换的圆形电路板以及传感器代表三种不同的动物——蛇、青蛙、海龟。

In an era where information transmission efficiency is paramount, vision has almost entirely constructed the reality of the digital age. The designer encourages people to rediscover the richness of sensory dimensions, paying more attention to the urban nature beyond electronic screens and pursuing a deeper sensory experience. By exchanging senses with animals, users are invited to experience familiar urban nature through non-human sensory perspectives. Three interchangeable circular circuit boards and sensors represent three different animals—snake, frog, and turtle.

树字延时
Treelapse

高世涵 | Fablab O Shanghai 数制工坊 | 2022
Shihan Gao | Fablab O Shanghai | 2022

树字延时是一种可以记录树的运动并将其转化为数字生成的延时摄影的装置。它可以绑在树上，内部的陀螺仪和加速计会检测并记录树的运动。生成的数据将保存到TF卡的日志文件中。然后，用户可以取出TF卡，使用日志文件生成数字艺术作品，其使用的算法与时间和收集到的数据密切相关。

Treelapse is a device designed to record the movement of trees and transform it into digitally generated time-lapse photography. It can be attached to a tree, where the internal gyroscope and accelerometer detect and record the tree's movements. The generated data is saved into a log file on a TF card. Users can then take out the TF card and use the log file to create digital art, with the artwork's algorithm intricately linked to the passage of time and the collected data.

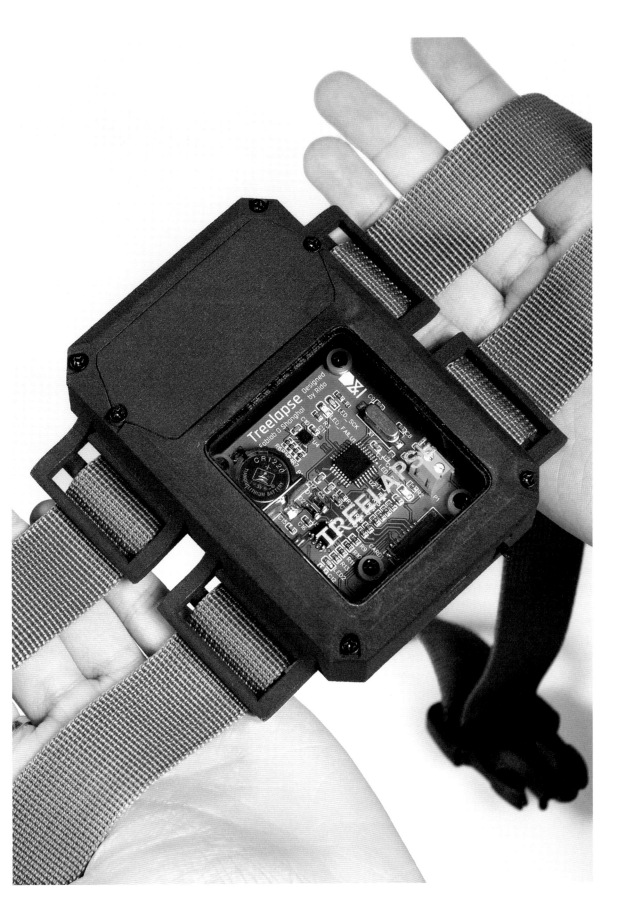

"小上海"智能板
Shanghaino: A PCB Board

萨维里奥·西利 | Fablab O Shanghai 数制工坊 | 2018
Saverio Silli | Fablab O Shanghai | 2018

由Fablab O Shanghaii数制工坊开发的"小上海"是世界知名的Arduino的定制版本。Arduino是一个易于使用的硬件和软件的开源电子平台。"小上海"是一个简单但功能强大的电路板。学生可以方便地理解其工作原理，了解各项组件，对其进行焊接，并上传第一个程序。

Shanghaino, developed by Fablab O Shanghai, is a custom version of the worldwide famous Arduino, an open-source electronics platform based on easy-to-use hardware and software. The Shanghaino is a simple but powerful PCB board. It allows students to learn how it works, understand its components, solder it and upload the first program on it.

AstroBot：基于开源硬件的创新天文观测设备
AstroBot: An Equatorial Mount Based on Open-source Hardware

高世涵 | Fablab O Shanghai 数制工坊 | 2022
Shihan Gao | Fablab O Shanghai | 2022

本作品是一款基于开源硬件的赤道仪产品，用户可以将其安装于自己的手机或是家用相机进行天文观测和拍摄。由于采用了与传统赤道仪不同的结构，AstroBot拥有更小的体积、可折叠的设计，为其提供了更强的便携性。其自带的校准、寻星和追踪程序大大降低了用户的使用门槛。AstroBot完全开源，用户可以自行获取三维模型、设计图和程序代码，并通过自己购买零件、3D打印来组装一个AstroBot。相比于价格高昂的传统赤道仪，AstroBot的制作成本不到1 500元，却能提供同样的性能和稳定性，以及更多的自动化功能。

The AstroBot is an equatorial mount product based on open-source hardware, allowing users to install their own smartphones or home cameras for astronomical observation and photography. By applying a structure different from traditional equatorial mounts, AstroBot offers a more compact size and a foldable design, providing enhanced portability. It features built-in calibration, star-finding, and tracking programs, significantly lowering the usage barrier for users. AstroBot is entirely open-source, enabling users to access 3D models, design drawings, and program code, and assemble their own AstroBot by purchasing parts and using 3D printing. Compared to expensive traditional equatorial mounts, AstroBot costs less than 1,500 yuan while offering the same performance and stability, along with additional automated features.

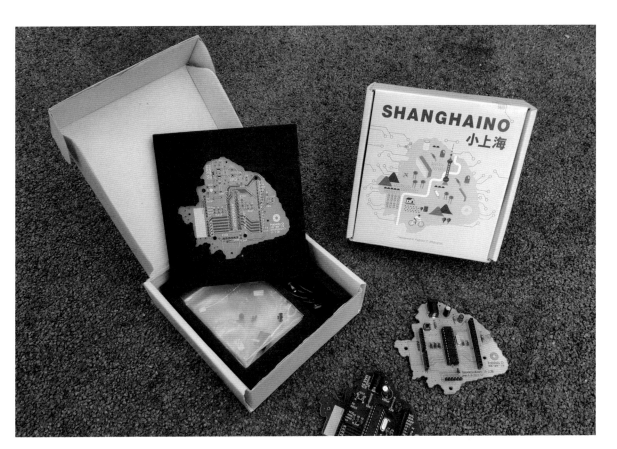
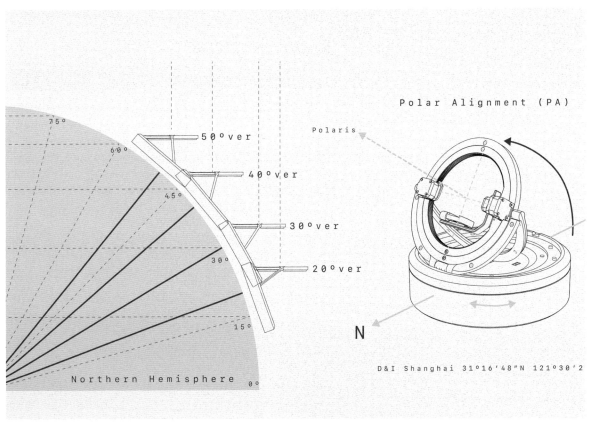

Ellissima 书柜
Ellissima

安东·西比克 | 2019
Aldo Cibic | 2019

Ellissima是一个由L形搭扣模块制成的书柜。作品采用了光滑的粉色漆木,固定在珍贵的老榆木底座上。这一彩色的倒置L形隔板的组合令人印象深刻,仿佛一座独具匠心的纪念碑。

Ellissima is a bookshelf made from L-shaped latch modules, crafted from smooth pink lacquered wood and set on a base of precious old elm. The impressive combination of colorful inverted L-shaped shelves creates a monument of unexpected harmony.

甜甜圈
Donut

安东·西比克 | 2019
Aldo Cibic | 2019

本作品是为坐着淋浴而设计的坐凳。之所以设计这个作品，设计师的说法是因为"懒"。甜甜圈，一个有趣而意想不到的圆形凳子，由实心桃花心木制成，中心有一个孔。这是一项不寻常的设计，亦是生活的好伴侣。

Donut is a stool designed for sitting while showering, created out of what the designer humorously describes as "laziness". This playful and unexpected round stool features a central hole and is made from solid mahogany. It is an unusual design, and serves as a companion in daily life.

咪咪、卟卟、噜噜
Mimi, Bubu, Lulu

安东·西比克 | 2019
Aldo Cibic | 2019

本作品由三张侧桌组成。其中，咪咪是黄色、粉红色和红色的漆木侧桌；卟卟是蓝色和红色的漆木侧桌；噜噜是浅绿色漆木侧桌，其圆柱形抽屉表面覆盖着金箔。

This work is consisted of three side tables. Mimi is a yellow, pink, and red lacquered wood side table; Bubu is a blue and red lacquered wood side table; Lulu is a light green lacquered wood side table with a cylindrical drawer surface covered in gold leaf.

回收陶瓷再生材料
Recycled Ceramic Materials

祎设计 | 2021—2023
Yi Design | 2021-2023

作为陶瓷产业不可降解的附属品,中国每年产生1 800万吨陶瓷废品,但大多数陶瓷生产城市没有适当的废物管理政策来负责任地解决这个问题,这些垃圾大多被非法倾倒或掩埋,且不能生物降解。因此,我们需要尽快找到废物管理的解决方案,祎设计正是由此而诞生的。目前祎设计已经开发完成了从原料到生产到应用的全链路创新流程,建立了大量的原材料库存,并校准了生产过程和运行能力,开发了一系列以废弃陶瓷为主要原料的再生制品。

与传统建筑材料相比,祎设计的陶瓷再生产品在降低碳排放的同时,提升了产品性能,生产过程中使用更少自然资源。这种新型技术所制造的产品不含水泥和胶等非环保黏合剂,可再次循环利用。

China generates 18 million tons of non-degradable ceramic waste annually, with most ceramic-producing cities lacking proper waste management policies. Much of this waste is illegally dumped or buried. Yi Design addresses this by creating a full-chain innovation process, from raw materials to production and application. The company has developed a robust inventory, calibrated production processes, and created a series of recycled products made primarily from discarded ceramics.

Yi Design's recycled ceramic products reduce carbon emissions, enhance performance, and use fewer natural resources compared to traditional materials. These new tech products do not contain cement or harmful adhesives and can be recycled again.

杨浦区未来生活原型街活力创新模拟平台
Yangpu District CityScope Living Line Innovation Simulation Platform

同济－麻省理工城市科学实验室，麻省理工学院媒体实验室城市科学组 | 2019
Tongji-MIT City Science Lab, MIT Media Lab City Science Group | 2019

CityScope Living Line 是为"NICE2035"设计的人机协同智能设计工具，服务于面向社区的生态系统设计。本项目通过Wi-Fi和摄像头采集数据，使用机器学习和计算机视觉技术进行分析，建立介入策略与创新活力的联系。以此模型开发算法，结合多媒介交互设计，为多利益主体构建友好、包容的沟通协作平台，用数据和实证辅助共同决策，为街区选择更理想的优化策略。

CityScope Living Line is a human-machine collaborative intelligent design tool created for the "NICE2035", aimed at designing community-oriented ecosystems. The project collects data through Wi-Fi and cameras, utilizing machine learning and computer vision technologies for analysis. It establishes connections between intervention strategies and innovative vitality. This model helps algorithm development, integrates multimedia interaction design, and creates a friendly and inclusive communication and collaboration platform for multiple stakeholders. It uses data and demonstrations to support collaborative decision-making, helping to select optimal strategies for the neighborhood.

NICE 网络
NICE Network

SustainX 可持续未来设计研究中心，NICE 社群 | 2015—2023
SustainX Lab, NICE Communities | 2015-2023

自2015年起，娄永琪教授牵头发起了NICE2035项目，旨在有意识地促进大学和社区的创新潜力相互碰撞、交融、协同共创，创立一种全新的"自下而上"的城市社区创新范式。在NICE相关实践中，挑选了十余个代表着NICE精神的"创新、创意、创业"项目，这些项目反映出了通过设计引领创新，加速大学知识溢出的潜力。NICE所孕育的创意社群，正透过设计教育的网络，从环同济经济圈扩散至上海这座城市，再延伸至国际高校联盟，展现着创新的无限可能。

Since 2015, Prof. Yongqi Lou has spearheaded the NICE2035 project, aiming to consciously foster the collision, integration, and co-creation of innovation potential between the university and communities, establishing a new "bottom-up" paradigm for urban community innovation. In the various NICE-related practices, we have selected over a dozen projects that embody the spirit of "innovation, creativity, and entrepreneurship". These projects reflect the potential of design-led innovation to accelerate the spillover of university knowledge. The creative community nurtured by NICE is spreading through a network of design education, expanding from the Tongji economic circle to the city of Shanghai, and further extending to an international alliance of universities, showcasing the limitless possibilities of innovation.

食物"魔"盒
Magic FoodBox

NICE 小灶，兜着走 | 2023
NICE Xiaozao, FooDZZ | 2023

"剩食"并非指吃剩的食物，而是指商铺内临过期或当天打烊前未售完的食物。将"剩食"作为"盲盒"售卖源自丹麦的"Too Good To Go"（弃之可惜的好东西）网站，入驻的商家能够在这里将店里的剩食以盲盒的形式打折出售，避免食物浪费。如今这一形式正在中国各大城市迅速扩张，那么就让我们看看大家收到的神秘盒子里都有些什么惊喜，或者惊吓？是真的在为减少食物浪费助力，还是又一场哗众取宠的闹剧？

"Surplus food" doesn't refer to leftovers, but rather to food items that are near expiration or unsold by closing time. The idea of selling surplus food as a "Magic Foodbox" originated from a Danish website called "Too Good To Go", where participating businesses can sell their surplus food at a discount in a box format, helping to prevent food waste. This concept is rapidly expanding across major cities in China. So, what surprises or shocks await in these boxes? Is this truly a step towards reducing food waste, or is it just another attention-grabbing stunt?

米的发酵
Fermentation of Rice

老乔 | 2021
Joyce | 2021

本作品通过米的发酵探索人与土地的关系。

This work explores the relationship between people and the land through the fermentation of rice.

插入大山里的苹果芯片
Inserting Chips via Apple into the Mountains

哈巴庄园 | 2023
Haba Manor | 2023

哈巴庄园位于海拔2600米的香格里拉三坝乡母支村的小盆地，高海拔的地理位置，相对平原离太阳更近，猛烈而悠长的日照赋予了苹果特有的"高原香气"。果园除了配有全程数字化管控，还装有元素级的水肥精准滴灌设备、灵活的环境控制系统，并且践行着国内最严格的检测标准。哈巴庄园还解决了附近村子里大部分纳西族青年人的就业问题，使他们无须走出大山，就可以在家门口自食其力。在一颗苹果里，我们能看见未来农业的可能性：同时连接人、土地与自然。

Haba Manor is located in a small basin in Muzhi Village, Sanba Township, Shangri-La, at an altitude of 2,600 meters. Benefiting from prolonged, intense sunlight, the apples develop a unique "plateau aroma". The orchard features fully digitalized management, precision drip irrigation, and flexible environmental controls, all while adhering to strict domestic testing standards. Haba Manor also provides local Naxi youth with employment opportunities, allowing them to work without leaving their mountain homes. Each apple embodies the future of agriculture, linking people, land, and nature.

气筑花园
Auradisus

高熙哲，刘奕扬 | JALAB 首饰实验室 | 2023
Xizhe Gao, Yiyang Liu | JALAB | 2023

这个装置借由人流行走的空气驱动气流。在快闪店周期内，利用这项可持续装置可以节约 7.2% 的空调能源消耗，并且降温过程中不产生任何温室气体。装置的核心部件利用了回收所得的香水瓶。设计师尝试对废弃的香水瓶进行全流程性的可持续改造，在引导顾客参与可持续行动的同时，同步完成香水瓶和快闪店展柜的完全回收和合理再利用。

This installation harnesses airflow generated by the movement of people to drive ventilation. It has achieved a 7.2% reduction in air conditioning energy consumption during the pop-up store's operation without emitting any greenhouse gases during the cooling process. The core components of the installation repurpose discarded perfume bottles. The designers attempted to implement a fully sustainable transformation of these bottles. While guiding customers to engage in sustainable actions, the project also ensures the complete recycling and rational reuse of the perfume bottles and pop-up store displays.

幸福的爱
Happy Love

赵世笺 | JALAB 首饰实验室 | 2023
Shijian Zhao | JALAB | 2023

"幸福的爱"系列是失蜡法铸造技术与透窗珐琅工艺制造的，代表七日神明的七款珐琅胸针作品集。

透窗珐琅是珐琅工艺中最复杂和优美的种类。低端市场倾向于使用普通填烧珐琅技艺，但年轻消费群体对珐琅接受程度尚处于上升阶段的情况下，透窗珐琅具有中高端市场潜力，能够为珐琅首饰市场注入新一轮活力。

"Happy Love" is a collection of seven enamel brooches representing "The Seven Deities", crafted using the lost-wax casting method and cloisonné enamel technique.

Cloisonné enamel is one of the most complex and beautiful types of enamel work. Nowadays the lower-end market tends to favor simpler enamel-filling, and younger consumers are increasingly embracing enamel jewelry, cloisonné enamel holds significant potential in the mid-to-high-end market. This collection aims to inject new vitality into the enamel jewelry market.

千花之海
Reborn of Miss Dior

高祎宁，吴淑婷 | JALAB 首饰实验室 | 2023
Yining Gao, Shuting Wu | JALAB | 2023

本作品使用废弃的迪奥小姐香水瓶，通过再造的方式，以展览装饰、吊灯、小灯的形式回到品牌方手中。作为展览装饰，作品可拆卸重组排列为不同图案；作为吊灯，可长久留在门店内；作为小灯，可成为赠与用户的礼品，甚至可作为一个新的产品副线进行生产销售。

Discarded Miss Dior perfume bottles are repurposed and returned to the brand in the form of exhibition decorations, chandeliers, and small lamps. As exhibition decorations, they can be disassembled and rearranged into different patterns; as chandeliers, they can remain in the store for long-term use; and as small lamps, they can be given as gifts to customers or even developed into a new product line for sale.

流沙的时间
Time of Sand

黄卉怡，王心怡 | JALAB 首饰实验室 | 2023
Huiyi Huang, Xinyi Wang | JALAB | 2023

本作品以回收迪奥的废弃香水瓶作为来源，构建起一个完整的玻璃回收循环系统。循环分为四个章节，探索了回收玻璃砂作为系列产品原料的生命力，并通过持续的创新推动材料的循环利用。同时，共创的活动形式也旨在引发参与者对自然资源消耗的反思，向参与者传达可持续的理念。

This work uses discarded Dior perfume bottles as a source to create a comprehensive glass recycling system. The cycle is divided into four phases, exploring the vitality of recycled glass sand as a raw material for a series of products. It also promotes material recycling through continuous innovation. As a result of collaborative activities, it also aims to provoke participants' reflections on the consumption of natural resources, conveying the ideas of sustainability.

再造 HDPE
HDPE Upcycling

李心瑜，汪淼 | JALAB 首饰实验室 | 2023
Xinyu Li, Miao Wang | JALAB | 2023

本作品是通过回收再造的化妆品护肤品包装HDPE材料，进行颗粒再造以及重塑新生的一个系列作品，产品形式多样，同时色彩绚丽，通过温度的不同调控以及色彩的不同组合，形成独一无二的回收再造时尚产品。

This series of works utilizes recycled HDPE materials from cosmetic and skincare packaging, transforming them into reconstituted pellets and reshaped into new, vibrant products. The products are diverse in form and rich in color, with unique combinations of heating techniques and color variations, resulting in one-of-a-kind recycled fashion items.

星云矿石
Newbula Mineral

钟以恒 | JALAB 首饰实验室 | 2022
Yiheng Zhong | JALAB | 2022

基于现有的知识，我们就能进行想象；但随着计算机技术的快速发展，想象力已不再是人类的专属。通过深度学习，机器可以更少地依赖人类来帮助它们了解这个世界。宇宙作为人类很少抵达的地方，我们对其的想象能力此时似乎和计算机处于同一个起点。设计师将星云、宇宙物质、机器想象和NFT这四个关键词结合起来，利用算法模型，去尝试探索认知的界限，想象和创造星云与神秘宇宙物质。

With existing knowledge, we can imagine; but with the rapid development of computer technology, imagination is no longer exclusive to humans. Through deep learning, machines can rely less on human input to understand the world. When it comes to the universe—an area scarcely reached by humans—our imaginative capacity seems to be on a par with that of computers. The designer combines the concepts of nebulae, cosmic matter, machine imagination, and NFTs, using algorithmic models to explore the boundaries of cognition, and to imagine and create nebulae and mysterious cosmic substances.

植年历
Plant Calendar

付蔚 | JALAB 首饰实验室 | 2021
Wei Fu | JALAB | 2021

"天地玄黄，宇宙洪荒。日月盈仄，辰宿列张。寒来暑往，秋收冬藏。"千字文的前六句道出了宇宙、星象、时间与农耕的联系。基于现代农耕科学的研究与实践，人们创造了农耕日历，让我们得以从植物生长节律的角度重新探索时间的形态。作品基于深度学习的生成式对抗网络（GAN）赋予了植物新的生长动态与种属特征，虚拟的花朵和果实具有部分自然界存在的原始形态，却又融合了机器对植物界的想象。结合花果特征为生成虚拟植物重新命名，定义其生长时节，制成植物年历，并通过后期手段重塑虚拟植物的生长过程，运用特效强化时间聚合、流转、涌动、逝去之态，让观者在人工智能想象的动态表盘墙中重新理解宇宙、时间、生命之意。

The opening lines of the "Thousand Character Classic" illuminate the connections between the universe, celestial phenomena, time, and agriculture. Building upon modern agricultural science, this project reimagines the concept of time through the growth rhythms of plants. Utilizing Generative Adversarial Networks(GAN), it generates virtual flowers and fruits that merge natural forms with machine imagination. Each virtual plant is renamed and assigned a growth season, forming a botanical calendar. Through post-processing and special effects, the dynamic growth processes of these plants are showcased, inviting viewers to reinterpret the essence of the universe, time, and life through the lens of artificial intelligence.

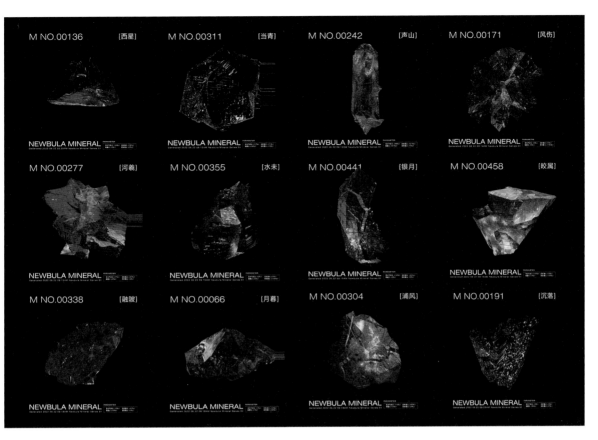

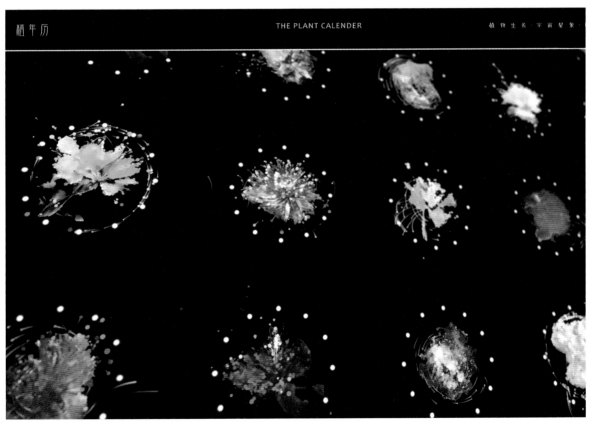

Introduction of Labs' Works

珍奇挚爱
Rare Love

赵世笺 | JALAB 首饰实验室 | 2023
Shijian Zhao | 2023

珍奇挚爱系列的设计灵感来自太湖石与珍珠的完美对比。作品以简洁的结构和数字化的造型技术展现了主题的精髓。太湖石造型的金属象征男性的坚韧，珍珠则象征女性的柔美，二者相拥，凸显出真爱相依之美。

The designer drew inspiration from the striking contrast between lake stones and pearls. The "Rare Love" series captures the essence of this theme through simple structures and digital modeling techniques. The lake-stone-shaped metal symbolizes masculinity, while the pearl femininity, highlighting the beauty of true love.

深度学习首饰
Newbula Mineral

郁新安 | JALAB 首饰实验室 | 2022
Xin'an Yu | JALAB | 2022

JALAB 首饰实验室主导的首个基于深度学习的首饰项目。本项目使用基于特征的生成对抗网络,在学习了 2 000 组以上特定风格的首饰作品之后,训练出首个针对戒指的生成神经网络。本项目的目的在于通过首饰图像的认知计算训练机器对于首饰的理解,最终实现基于特定风格首饰的自主生成。

This is the first jewelry project led by JALAB that is based on deep learning. The collection utilizes feature-based Generative Adversarial Networks (GANs) that was trained on over 2,000 jewelry designs of specific styles to develop the first neural network focused on generating rings. The goal of the project is to train the machine's understanding of jewelry through cognitive computing of jewelry images, ultimately enabling the autonomous generation of jewelry in specific styles.

咖啡土壤
Land of Coffee

周洪涛 | 造物实验室 | 2022
Hongtao Zhou | Making Lab | 2022

使用黏合剂将咖啡渣层层堆叠来祝福新庄园的诞生，代表了咖啡种植庄园景观的不同层次。

By layering coffee grounds with adhesive, this piece celebrates the birth of a new coffee plantation, symbolizing the various strata of a coffee-growing landscape.

咖啡之环
Coffee Rings

周洪涛 | 造物实验室 | 2022
Hongtao Zhou | Making Lab | 2022

这些咖啡圈之间的联网暗喻了我们在家里的日常生活、玩耍和工作时，不经意间遗留在桌上的咖啡印渍。

The interconnected coffee rings subtly symbolize the traces of coffee stains we unconsciously leave on tables during our daily activities, such as playing, working.

咖啡云托盘
Coffee Cloud Plate

莫娇 | 材料与应用创新实验室 | 2021
Jiao Mo | Material and Application Design Laboratory | 2021

咖啡杯、餐盘、蛋糕……咖啡桌和咖啡店里总是充满了圆形的东西，为什么我们的托盘不是圆的？托盘上要放很多的圆！本作品是为一餐丰盛的下午茶设计的。深色的纤维板中使用了回收来的咖啡渣，其散发出的香气时时为咖啡时间代言。

Coffee cups, plates, cakes... Coffee tables and cafés are always filled with round objects. So why isn't our tray round? The dark fiberboard is made from recycled coffee grounds. The aroma it carries perfectly repre-sents coffee time.

咖啡渣音响
Coffee Cloud Plate

颜呈勋，曾雪婷，吕丹丹 | 2020
Bill Yan, Xueting Zeng, Dandan Lyu | 2020

设计师对咖啡渣板进行了性能分析总结：有吸音性、作为表面材料有独特美感。他们认为咖啡渣板是一款适合做音响的面板材料。

The designer conducted a performance analysis of coffee ground fiberboard and found it to have sound-absorbing properties and a unique aesthetic as a surface material. Therefore, they concluded that coffee ground fiberboard is an ideal material for speaker panels.

插接式手机座
Phone Stand

造物实验室，材料与应用创新实验室 | 2020
Making Lab, Material and Application Design Laboratory | 2020

科思创"无醛咖啡渣循环再生板"完美展现了无甲醛聚氨酯板材和咖啡渣循环利用的解决方案，环保且有助于降低建筑能耗和碳排放。支架结构源自中国传统榫卯三交栿，三根木条紧密结合，展现出传统建筑智慧的坚固与美观。作品象征科思创与客户合作共赢的精神。

The Covestro "Formaldehyde-Free Coffee Ground Recycled Board" showcases polyurethane's formaldehyde-free material and sustainable coffee ground reuse, improving environmental quality and reducing energy and carbon emissions. Inspired by traditional Chinese Sanjiao Cheng mortise and tenon, the stand's interlocking wooden strips demonstrate structural integrity and aesthetic beauty, symbolizing Covestro's cooperative spirit with its customers.

Introduction of Labs' Works

树桌
Tree Table

莫娇 | 材料与应用创新实验室 | 2021
Jiao Mo | Material and Application Design Laboratory | 2021

对于熟悉CAD的设计师来说，这张边桌的造型来自软件中树的平立面图标，在设计方案中种树是丰富视觉效果的绝佳手段，也隐含着人们对自然的美好向往。本作品运用的几乎是同样的数字文件，以数控加工的方法切割后插接制作而成，在概念上把"种树"的电子行为转变为物化行为。采用循环咖啡渣纤维板与胶合板夹心复合板材，在保证家具结构强度的同时，充分展现循环咖啡渣纤维板的表面肌理带来的自然质感。

For designers familiar with CAD, the shape of this side table is inspired by the tree icons used in software's elevation and plan views. In design proposals, adding trees has been a beautiful way to enrich the visual effect and subtly instill people's appreciation for nature. Using nearly the same digital files, the table is created by CNC cutting and interlocking the pieces, conceptually transforming the electronic act of "planting trees" into a tangible one. The table is made from a composite material of recycled coffee ground fiberboard and plywood, ensuring structural strength while fully showcasing the natural texture of the recycled coffee ground fiberboard.

萌虎凳
Tiger Stools

邢岩 | 材料与应用创新实验室 | 2020
Yan Xing | Material and Application Design Laboratory | 2020

作品采用儿童玩具与榫卯结合的构思，趣味好玩，背后是一颗能够长大的孩子心。作品用简化的榫卯结构在增加结构稳定性的同时，传统榫卯结构和板材插接也体现出独有情趣。咖啡渣循环再生板材采用数控车铣精密加工，确保精度的同时，避免板材浪费，从而提高板材利用率，与环保、可持续发展的理念高度契合。

These pieces combine the concept of children's toys with traditional mortise and tenon joints, resulting in a playful and engaging design that reflects a child's ever-growing imagination. The simplified mortise and tenon structure enhances stability, while the use of traditional design elements and board interlocking adds a unique charm. The stools are made from recycled coffee ground boards, precisely machined using CNC milling, ensuring accuracy and minimizing material waste. This approach not only improves material utilization but also aligns with the principles of environmental sustainability.

Introduction of Labs' Works

文字景观
Textscape

周洪涛 | 造物实验室 | 2021
Hongtao Zhou | Making Lab | 2021

本项目重新强调了"印刷"一词在现代技术"3D打印"中的新内涵。文字景观重新在三维界面打印文字，生成三维的文字文档，并利用高低起伏直观地展示文本的视觉主题，形成城市、景观或人物。本项目使内容与形态完美结合，促进了文化与形态的认知交互，形成了城市礼物设计。目前已经形成近百件的系列产品形态，并在从产品到公共环境的不同尺度的领域得以应用。

Textscape redefines the meaning of "printing" within the context of modern "3D printing" technology. This project involves printing text in a three-dimensional interface, generating 3D textual documents, and visually representing the text's theme through varying elevations. The result is the creation of urban landscapes, scenes, or figures that blend content with form, fostering cognitive interaction between culture and shape. This approach has led to the development of nearly a hundred products, which have been applied across various scales, from individual products to public environments, making them ideal urban gifts.

能量木
Energy Wood

周洪涛 | 造物实验室 | 2020
Hongtao Zhou | Making Lab | 2020

再生木材通过复杂的切割和弯曲塑造出不同的造型。在室温和标准湿度下，设计师将这些细长的木材弯曲，形成抽象的咖啡树叶纹理。

Recycled wood is intricately cut and bent into various forms. Under room temperature and standard humidity, these slender strips of wood fibers were bent, forming abstract coffee leaf textures.

犀皂
Cup Soap

章雨晨，肖雨晴 | 材料与应用创新实验室 | 2020
Yuchen Zhang, Yuqing Xiao | Material and Application Design Laboratory | 2020

本作品基于文玩隐形分级系统优化的服务设计。作品将肥皂"洗手"的功能，和文玩爱好者"盘玩"的日常动作相融汇，并与保护犀牛的主题巧妙结合，寓意"盘"在掌心，爱在心头。

This design is based on an optimized service model using an invisible grading system for collectible cultural artifacts. It creatively combines the function of soap for "washing hands" with the habitual "rubbing" motion of collectors, aligning it with the theme of rhino conservation. The act of "rubbing" the soap in one's palm symbolizes care and love for the rhino, connecting the tactile experience with a deeper conservation message.

嘻犀、想象、独角鲸
Icero, Elephane, Narwhal

赵韵景，祝思烨，麻珮琦 | 材料与应用创新实验室 | 2020
Yunjing Zhao, Siye Zhu, Peiqi Ma | Material and Application Design Laboratory | 2020

几年前新冠疫情的突然暴发可以说是环境对人类发出的警告，后新冠时代的新常态让人类更多地关注自然，体会自然与人的和谐共处的重要性。从南非每天被猎杀的犀牛，到迷失北上的西双版纳象群，还有离我们很远，却在北冰洋里的继续被散弹捕猎的独角鲸，都是我们生态圈中不可缺少的一分子。嘻犀、想象、独角鲸产品通过木纤维复合塑料（Formi, UPM）制成，以珍稀野生动物——犀牛、象、独角鲸为原型，设计成为一套供涂鸦用的玩具，配合丙烯彩笔的着色性能和优良的颜色覆盖性能，提供可以反复涂鸦创造的乐趣。与此同时，产品能够激发青少年儿童对自然的热爱，增强对动物保护的意识，发挥想象力创造力和美育教育。

The sudden outbreak of COVID-19 a few years ago can be seen as a warning from the environment to humanity. The post-pandemic new normal has led people to pay more attention to nature and understand the importance of harmonious coexistence between humans and the natural world. The daily killing of two rhinos in South Africa, the wandering elephants of Xishuangbanna, and the narwhals in the Arctic, hunted from afar, are all indispensable members of our ecosystem. The Icero, Elephane, Narwhal product series is made from wood fiber composite plastic (Formi, UPM) and designed as a set of graffiti toys modeled after endangered wildlife—rhinos, elephants, and narwhals. These toys are paired with acrylic markers, offering excellent coloring and coverage for repeated creative expression. The series aims to inspire a love for nature, raise awareness of wildlife conservation, and foster imagination, creativity, and aesthetic education in children and teenagers.

Introduction of Labs' Works

社交距离——城市公共家具设计
Social Distance

朱曦 | 材料与应用创新实验室 | 2021
Xi Zhu | Material and Application Design Laboratory | 2021

"宅"不仅是一种生活状态,而且已慢慢成为人们的思维方式。宅的思维已走出家门,延伸至户外。社交距离成为人们对自我安全的基本心理设定。以往常用的面对型、直线型公共座椅的使用方式面临挑战。本作品意在用更少的公共空间建立更宽松的个人"宅"空间,通过一体化的造型语言,表达在保持安全社交距离的同时,我们可以靠得更近。三人座的城市公共座椅设计,可以选择更多的配色融入环境。

"Staying at home" has become not just a lifestyle but a mindset that has gradually extended beyond the home to outdoor spaces. The concept of social distancing has become a basic psychological setting for personal safety. Traditional face-to-face or linear public seating arrangements are now being challenged. This design aims to create a more relaxed personal "home" space using less public space. Through an integrated design language, it expresses the idea that while maintaining safe social distances, we can still come closer together. The three-person urban public seating design offers various color options to fit in the environment.

花鼓新传系列家具
Penta-Circle Series

莫娇 | 材料与应用创新实验室 | 2023
Jiao Mo | Material and Application Design Laboratory | 2023

本系列作品系中式五足家具,具有沉稳雄浑的风格,以更具变化的环形曲面,借由拱形的支撑强度,呈现抽象干练的造型,以突出座面数字生成装饰风格的千变万化,超以象外,得其环中。五足是中式家具中的一种独特结构,独树一帜。如明式家具中的五开光花鼓凳、五足海棠桌等。五足制式设计之妙在于平稳中汇聚俏丽,打破了通常所见的四足家具的四平八稳、刻板划一,而五足家具则显得灵动活泼、别具一格。

The series is rooted in the traditional Chinese five-legged furniture design, featuring a more dynamic circular surface and robust and majestic style. Utilizing the strength of arched supports, the design presents an abstract and refined form, highlighting the endless variations in the digitally generated decorative style of the seat surface. The five-legged structure is a unique feature of Chinese furniture, standing out with its distinctive character, as seen in classic pieces like the Ming-style five-legged flower drum stools and begonia tables. The brilliance of the five-legged design lies in its ability to combine stability with elegance, breaking away from the often rigid and uniform four-legged furniture. In contrast, five-legged furniture appears lively and dynamic.

组子系列墙饰
Kumiko Coaster, Wall Decor

谢雅迪 | 造物实验室 | 2020
Yadi Xie | Making Lab | 2020

组子是传统且独特的日式木工工艺,起源于公元8世纪并延续至今。组子的制作过程主要是将零件精密仔细计算与切割雕刻,促成木件之间通过压力相互紧密连接,而不用一滴胶水或一颗钉子,打造出独特的图案花纹。各式组子图案常用于家具门窗装饰,大小不一。在设计师的后续的研究中,将会出现新的图案及组成,并且带有新的意义。未来,设计师将会持续探索新材料组子的可能性与加工工艺,以及这项传统工艺在可持续设计中的新角色。

Kumiko is a traditional and unique Japanese woodworking technique that originated in the 8th century and continues to be practiced today. The process involves precisely calculating, cutting, and carving components to fit them tightly together through pressure alone, without using any glue or nails, resulting in intricate patterns. Kumiko designs are often used in furniture, doors, and window decorations, varying in size. In the designer's ongoing research, new patterns and compositions will emerge, with new meanings. In the future, the designer will focus on the possibilities of using new materials for Kumiko and its processing techniques, as well as its evolving.

Introduction of Labs' Works

创意的可计算性
The Computability of Creativity

范凌 | 设计人工智能实验室 | 2017—2023
Ling Fan | Design A.I. Lab | 2017-2023

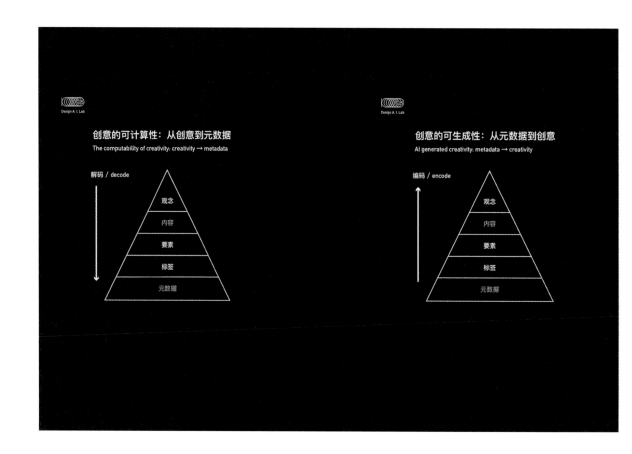

创意的可计算性探讨如何建立创意数据基础框架，使创意通过数据得到描述。我们搭建了不同类型的创意数据集和知识图谱，用数据进行创意文化研究。这是数字人文的一个分支，也为我们帮助机器理解创意提供了基础。

The Computability of Creativity explores how to establish a data-based framework for describing creativity through data. The project involves building various types of creative datasets and knowledge graphs, using data to conduct research in creative culture. This work represents a branch of digital humanities and lays the foundation for helping machines understand creativity.

AIGC 创客松作品
The Works of AIGC Creatorthon

AIGC 创客松的创作者们 | 设计人工智能实验室 | 2023
The Creators of AIGC Creatorthon | Design A.I. Lab | 2023

"AIGC创客松"鼓励参与者手脑并用，发挥创意，希望通过实践让每个人都可用AI绘画，成为创作者。AIGC工作坊使用的工具"MuseAI"由特赞科技（musai.cc）提供。本页所有图片都由相关提示词通过AI生成。

The AIGC Creatorthon encourages participants to engage both mind and hand, fostering creativity with the goal of empowering everyone to become creators through AI-driven art. The workshops utilized the tool "MuseAI", provided by Tezign Technology (musai.cc). All images on this page were generated by AI using the relevant prompts.

排列设计和深度排列设计
Permutation Design & Deep Permutation Design

康思大 | 尚想实验室 | 2015—2023
Kostas Terzidis | Shang Xiang Lab | 2015-2023

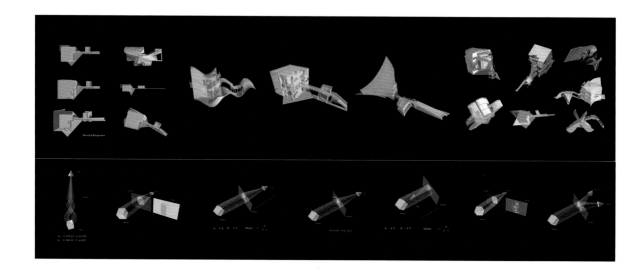

编码是一种强大的工具，不仅是因为它让学生能够以更系统的方式表达、解决和执行他们的想法，而且最重要的是，它让人们得以探索超越想象的新想法。程序员常常对自己的代码感到惊讶——与他们想象的相比，结果更好、不同或不可预测。排列设计是指在复杂形式的合成中使用元素的详尽组合的设计方法，而深度排列设计则基于排列和神经网络探索人工智能的设计潜力。

Coding is a powerful tool, not only because it allows students to express, solve, and execute their ideas more systematically, but more importantly, it enables the exploration of new ideas beyond imagination. An intriguing fact is that programmers are often surprised by their own code—the results can be better, different, or unpredictable compared to what they initially envisioned. Permutation design refers to a design approach that uses exhaustive combinations of elements in the synthesis of complex forms. Deep permutation design takes this further by exploring the potential of artificial intelligence in design through the use of permutations and neural networks.

折面动态艺术装置 04
Mesh Motion #04

张周捷 | 艺术与人工智能实验室 | 2020
Zhoujie Zhang | Art & Artificial Intelligence Lab | 2020

如今城市化进程加快，艺术家张周捷希望通过该作品唤起人们对自然的感知，使观者重新思考人与自然的关系。这件翼状互动装置依据其在地性，全面融合鹤的形意与装置机械原埋，通过计算机算法这样一个代表科技和未来的前瞻性创作媒介，将艺术家对于自然的诗性感知与发现体现于作品之中。

Amid accelerated urbanization process, artist Zhoujie Zhang hopes to evoke a sense of nature in viewers through this work, prompting them to reconsider the relationship between humans and nature. This wing-shaped interactive installation is site-specific, fully integrating the form and essence of a crane with mechanical principles. Utilizing computer algorithms, the artist places his poetic perception and discovery of nature in the work.

数制蓝染
Indicode

彭月儿，程诺，苏梓裔，陈罗千绘，肖明阳，陆思渊，汪淼 | 载运工具与系统创新设计实验室 | 2023

Yue'er Peng, Nuo Cheng, Ziyi Su, Luoqianhui Chen, Mingyang Xiao, Siyuan Lu, Miao Wang | Next Mobility Lab | 2023

从秦汉时期到数字化时代，蓝染工艺历经千年岁月沉淀，未来手工艺如何发展？传统手工艺工序复杂，普通人难以触及。本作品遵循蓝染的传统工序，通过数字技术助力每个人定制蓝染作品，打造蓝染数字制作平台，驱动蓝染手工艺的新想象，用更个性化、便捷的方式创造蓝染工艺品。

From the Qin and Han dynasties to the digital age, indigo dyeing has undergone millennia of refinement. How will this traditional craft evolve in the future? Traditional indigo dyeing involves complex processes that are difficult to access. This project follows these traditional steps while leveraging digital technology to enable everyone to customize their own indigo-dyed works. By creating a digital platform for indigo dyeing, this project offers new possibilities for the craft, allowing for the creation of indigo art in a more personalized and convenient way.

肌理之音
Sound of Texture

王丹，高祎宁，刘泊岩，潘贵玲 | 载运工具与系统创新设计实验室 | 2023

Dan Wang, Yining Gao, Boyan Liu, Guiling Pan | Next Mobility Lab | 2023

土壤、年轮、树皮、叶脉，自然界万物都生长着各种各样的肌理。在深入研究之后，设计师们发现这些表面肌理蕴含了丰富的信息，他们希望通过现代化数字手段对自然界的肌理进行提取与转译，去构建一个"人肌交互"的世界。从输入端转换到输出，从自然材料转换到人造物，在整个系统中，设计师们希望通过算法让人造环境认识自然肌理，完成从自然肌理到声音场景的转换。

Soil, tree rings, bark, and leaf veins—various textures are found throughout nature. Through in-depth research, the designers discovered that these surface textures contain rich information. Their goal is to use modern digital methods to extract and translate these natural textures, creating a "human-texture interaction" world. From input to output, and from natural materials to man-made objects, their system aims to enable artificial environments to recognize natural textures, ultimately transforming natural textures into soundscapes through algorithms.

Introduction of Labs' Works

碳知硅应
Project Si by C

彭月儿，程诺，苏梓裔，陈罗千绘，肖明阳，陆思渊，汪淼 | 载运工具与系统创新设计实验室 | 2023

Yue'er Peng, Nuo Cheng, Ziyi Su, Luoqianhui Chen, Mingyang Xiao, Siyuan Lu, Miao Wang | Next Mobility Lab | 2023

人类从取材自然开始到效仿自然，构建了现在的人造世界。未来我们如何造物？自然生物为了适应万千生境，进化出了各具特点的响应机制以保护自身。本作品依然取材于自然，但保留了生物活性，区别于仿生而倡导碳硅融合，旨在探究生物感知触发人造响应的未来式，以更自然、更灵敏的方式感知世界，服务未来生活。

Humanity has evolved from using natural materials to mimicking nature, creating the artificial world we inhabit today. What will we create in the future? Natural organisms have evolved unique response mechanisms to adapt and protect themselves in various environments. This project still draws material from nature but emphasize the integration of carbon and silicon, preserving biological activity. The focus is on exploring future-oriented artificial responses triggered by biological perceptions, aiming to perceive the world in a more natural and sensitive way and serve future living environments.

心游漫旅
Mind Wandering

吴淑婷，王自源，王怡然，顾天晨 | 载运工具与系统创新设计实验室 | 2023

Shuting Wu, Ziyuan Wang, Yiran Wang, Tianchen Gu | Next Mobility Lab | 2023

在能量、信息、人员高速流动的时代，你多久没有在旅途中安静闲适，放空身心了？生活繁忙的穿行中，我们仍渴望寻觅一场心流回归之旅。在这段旅程中，设计师们更关注触觉体验的正念意义。以皮肤触觉介入，他们发现借由两点阈度量个体触觉空间的辨别能力，通过相应调控振动模块，让用户在旅途中进入心流状态，进而让触觉成为"知"的器官，追寻感知合一，以触感唤起流动中的心游之旅。

In an era of rapid flow of energy, information, and people, how long has it been since you experienced a moment of tranquility and relaxation during your travels? Amid the bustling pace of life, we yearn for a journey of mindful return. This journey focuses on the mindfulness of tactile experiences. By incorporating skin touch and measuring the two-point threshold of individual tactile space, the designers use vibration modules to help users enter a state of flow during their travels. This allows touch to become an organ of knowledge, merging perception and sensation to evoke a journey of heart and flow through tactile experiences.

设计生态集锦展区
Design Ecological Exhibition Area

1000 个盒子的 365 天之旅
1000 Boxes' 365-Day Journey

同济大学设计创意学院，SustainX 可持续未来设计研究中心 | 2022
College of Design and Innovation at Tongji University, SustainX Lab | 2022

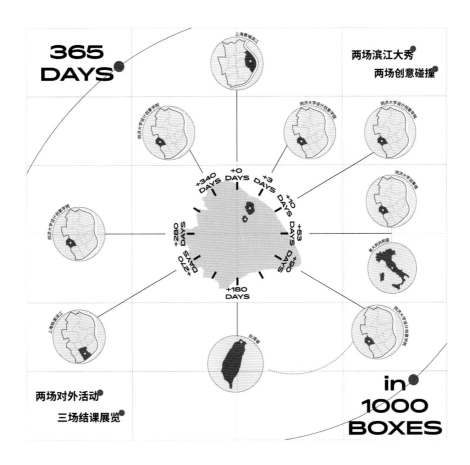

2022年首届WDCC主题展中，近1 000个红酒箱作为基础模块，被拼接组合成各种形式的展台。展览结束后，这些箱子又回到同济大学设计创意学院，积极参与毕业设计展、专业课程展览、青少年创意设计大赛、海峡两岸创客大赛等活动，承载了无数创意思维的碰撞。这些红酒箱不再仅仅是一次性的装置，它们延续了首届WDCC主题的理念，成为可持续设计的象征。红酒箱的接力不仅是WDCC文化的传播，也展现了设计社群的凝聚力，激发起更多人为可持续发展而设计的思考。

In the inaugural WDCC theme exhibition of 2022, nearly 1,000 wine boxes served as fundamental modules, creatively assembled into various forms of display stands. After the exhibition, these boxes returned to College of Design and Innovation at Tongji University, actively participating in graduation design exhibitions, professional course exhibitions, youth creative design competitions, and cross-strait maker competitions, and other events. They became vessels for countless creative exchanges. These wine boxes are no longer mere temporary installations; they have continued the theme of the inaugural WDCC, becoming symbols of sustainable design. The journey of these wine boxes represents not only the dissemination of WDCC culture but also showcases the cohesion of the design community, inspiring further thoughts on designing for sustainability.

同平共振
Resonance Between Tongji and Siping

苏雅默 | 2023
Jarmo Suominen | 2023

同济大学设计创意学院外籍教授,来自芬兰的苏雅默教授在展览开幕的现场,即兴创作了这幅巨幅手绘白描壁画。作品描绘了与杨浦区四平路街道合作的散布于设计创意学院周边社区的实验室群落的日常生活场景。娄永琪教授以"同(济)(四)平共振"(四平路街道金晔主任用语)为题,为作品撰写了"城校共创,三区联动,凝聚创意社群;设计驱动,创造需求,倒逼创新转化"的题签。

At the exhibition opening, Prof. Jarmo Suominen from Finland, a visiting professor at College of Design and Innovation, Tongji University, created an impromptu large-scale hand-drawn mural. The mural depicts the daily life scenes of laboratory clusters scattered around the community of the College of Design and Innovation in collaboration with the Siping authorities. Prof. Yongqi Lou wrote a thematic inscription for the artwork titled "Resonance Between Tongji and Siping".

开 / 盖
Open-Close

辛思想设计思维与战略咨询 | 2023
XXY Innovation | 2023

这是一个关于瓶盖，而不止于瓶盖的展览。瓶盖在生活中过于常见，我们往往对"开/盖"简单动作习以为常，却从未注意过它们并不简单的设计。其实，正是它们趋于完美、适应人类行为的设计使我们对它们习以为常。每一个被使用的瓶盖背后隐藏的都是一个精巧的小工程。"开盖"展览旨在启发人们关注生活中的细节。

This exhibition is about bottle caps, and much more than that. Bottle caps are so common in our lives that we often take the simple act of "opening/closing" for granted, without noticing the intricate design behind them. In reality, it is their near-perfect design, tailored to human behavior, that makes them so familiar. Each used bottle cap hides a small, ingenious engineering feat. The "Open-Close" exhibition aims to inspire people to pay attention to the details in everyday life.

HOTO 小猴工具箱
HOTO Toolbox

HOTO 小猴 | 2023
HOTO | 2023

HOTO小猴工具箱系列，是集合HOTO小猴家庭工具的新式工具箱，根据不同用户需求，提供多款不同工具的组合以供选择。内部的每件工具有秩序地紧凑排列，并拥有统一的设计语言。每件工具融入协力操作的巧思，使得工具更加直观易用，降低了使用门槛，使它们成为人人适用的工具箱。浅色系干净舒适的外观视觉，传达给人们工具箱可以更优雅、日常化的暗示。"咔嗒"一声，隐藏式蝴蝶卡扣设计，给予工具箱顺滑的开箱体验，让使用者的创作愉快开始。

The HOTO toolbox series represents a new kind of toolbox that consolidates HOTO's range of household tools. Offering various tool combinations to meet different user needs, each tool is compactly arranged within the toolbox and features a unified design language. The tools are designed with collaborative operation in mind, making them more intuitive and easier to use, thereby lowering the entry threshold and making them accessible to everyone. The clean and comfortable light-colored exterior suggests that toolboxes can be both elegant and everyday. With a satisfying "click", the concealed butterfly clasp design provides a smooth unboxing experience, setting the stage for a delightful creative process.

莫比斯环椅
Möbius Ring Chair

杨明洁 | 2022
Mingjie Yang | 2022

手工艺的自由与个性在数字时代以另一种方式得以体现，但结果并不相同。这一点在采用回收材料与FDM 3D打印成型工艺完成的羊舍（杨明洁创立的品牌）莫比斯环椅中有所体现，类似莫比斯环多维空间转折结构是工业时代无法实现的。在打印过程中，控制喷头的路径，尝试了一种不确定的手工艺质感，最终呈现的是一种属于数字时代的美学特征，回收材料的运用也符合可持续发展的设计原则。

The freedom and individuality of craftsmanship are expressed in a new way in the digital age, though the outcomes differ. This is evident in the Möbius Ring chair(YANG HOUSE, a brand founded by Mingjie Yang) crafted from recycled materials and formed using FDM 3D printing technology. The Möbius ring's multidimensional spatial twists represent a structural complexity unattainable in the industrial era. During the printing process, controlling the printhead's path introduced a handcrafted texture with an element of uncertainty. The result is an aesthetic characteristic unique to the digital age, with the use of recycled materials aligning with sustainable design principles.

纸剃刀
Paper Razor

贝印株式会社 | 中国设计智造大奖铜奖作品 | 2022
Kai Corporation | DIA BRONZE | 2022

本产品是一款采用纸柄和金属头的剃须刀，可以减少塑料的使用。这款剃须刀采用扁平包装，组装后即可使用，厚度仅为3mm，重量仅为4g。产品上的图案可以替换。本设计为用户提供了情绪价值，让他们有了选择的乐趣，在剃须的基本功能之外增加了额外的愉悦感。

A shaver designed with a paper handle and metal head to reduce plastic use. It is packed flat and assembles for use, with a thickness of just 3 mm and a weight of just 4 g. The printed graphic can be changed to create a variety of patterns. This is a new concept that provides the emotional value of the enjoyment of selection and the enjoyment of use in addition to the basic function of shaving.

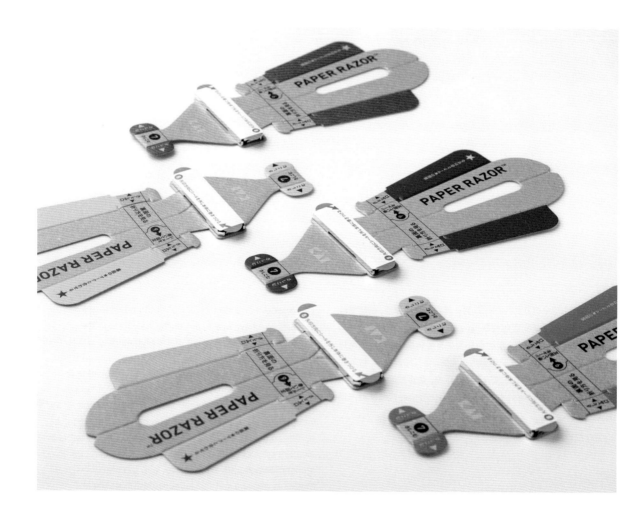

卡尔佐内
Calzone

卡尔佐内原指意大利一种半圆形馅饼，本产品采用了与此类似的外观，是一种由硅胶制成的可重复使用的折叠板，可用于取代传统的一次性盘子和塑料拉链袋等一次性物品。去野餐或徒步旅行时，它可以用作盘子或替代塑料拉链袋存放零食。此外，它还可以折叠起来存放剩余的食物。它由医用级液体硅胶制成，可以安全地用于微波炉、烤箱加热和冷冻储存。卡尔佐内将成为一种可持续产品，即使多次使用也不会损坏，对人体无害，方便，并且可以与附件一起长期使用。

BDCI 设计工作室 | 中国设计智造大奖佳作奖作品 | 2022
BDCI | DIA HONORABLE MENTION | 2022

Calzone is a reusable folding plate made of silicone. It can be used as dishware, replacing disposable items such as conventional disposable plates and plastic zipper bags. When going on picnics or hiking, it can be used as a plate or to store snacks instead of zipper bags. Also, it can be folded to store leftover food. As Calzone is made of medical-grade liquid silicone, it is safe to microwave, oven-heat, and store frozen. Calzone will become a sustainable product with a design that is not damaged even if used many times, is harmless to the human body, is convenient, and can be used for a long time with attachment.

欧姆龙喘鸣检测仪
OMRON WheezeScan

欧姆龙健康医疗事业株式会社 | 中国设计智造大奖铜奖作品 | 2021
OMRON Healthcare Co., Ltd. | DIA BRONZE | 2021

本产品可用于检查幼儿的喘息情况，并在哮喘症状急性发作之前进行治疗。配套的哮喘日记应用程序记录每日症状和处方药物。当与记录喘息发生、症状和药物一起使用时，本系统可以准确地与医生共享家庭哮喘管理信息。

WheezeScan checks for wheezing in small children and enables treatment before asthma symptoms become an acute attack. The Asthma Diary application records daily symptoms and prescribed medication. When used together to record wheezing occurrences, symptoms, and medications, the system accurately shares home asthma management information with a doctor.

FabLittleBag：经期废物处理袋
FabLittleBag: Disposal Bag for Period Product Waste

Loopeeze 有限公司 | 中国设计智造大奖佳作奖作品 | 2021
Loopeeze Ltd trading as FabLittleBag | DIA HONORABLE MENTION | 2021

本产品来源可持续，并已获得专利。它适用于卫生棉条、护垫和避孕套的一次性处理，易于单手使用，不透明，密封性好，干净卫生。它能有效防止经期废物流入河流、海洋和海滩，免去了用户丢弃时的烦恼与尴尬。

This product is sustainably sourced, patented, fit for purpose disposal bag for tampons, pads and condoms that's easy to use one handed, opaque for discretion, and seals closed for good hygiene. It keeps period product waste out of the rivers, oceans and beaches. It takes the stress and the mess out of binning.

Hable One：智能手机盲文键盘
Hable One: Braille Keyboard for Smartphones

Hable 公司 | 中国设计智造大奖佳作奖作品 | 2021
Hable One B.V. | DIA HONORABLE MENTION | 2021

Hable是一家位于埃因霍温的社会影响力初创企业，致力于识别并解决视障人士在使用智能手机、电脑等时面临的问题。Hable One是一款用于打字和操作智能手机的通用无线控制器，使视障人士能够独立获取信息。

Hable is an Eindhoven based social impact start-up. They identify and solve problems visually impaired face while using technology – smartphones, computers etc. Hable One is a universal wireless controller for typing and operating smartphones, which enables visually impaired to gain access to information and gain independence.

Scribits：绘画机器人
Scribits: A Write and Erase Robot

卡洛·拉蒂事务所 | 中国设计智造大奖银奖作品 | 2020
Carlo Ratti Associati | DIA SILVER | 2020

Scribit由麻省理工学院教授卡洛·拉蒂领导的屡获殊荣的设计和创新公司卡洛·拉蒂事务所设计，是一款书写和擦除机器人，可以将任何垂直表面变成低刷新屏幕，在其上显示来自网络、用户的信息。作为"墙壁打印机"，Scribit开创了一种呈现数字内容的新方式，并允许用户立即重新配置和个性化垂直平面。继2018年成为Kickstarter和Indiegogo上最成功的全球众筹活动之一后，Scribit的融资总额超过240万美元。2019年，Scribit荣获产品设计红点奖，并被《时代》杂志评选为年度100项最佳发明之一。

Designed by Carlo Ratti Associati, the award-winning design and innovation firm led by MIT Prof. Carlo Ratti, Scribit is a write and erase robot that can turn any vertical surface into a low-refresh screen on which to display information from the web, user-generated content and art. Functioning as a "printer for walls", Scribit ushers in a new way of presenting digital content and allows the user to instantly reconfigure and personalize a vertical plane. Following one of the most successful global crowdfunding campaigns on Kickstarter and Indiegogo in 2018, the vertical plotter totalized over 2.4 million USD. In 2019 Scribit was awarded RedDot for Product design and singled out as one of the 100 best inventions for the year by *TIME* Magazine.

Petit Pli：可生长服装
Petit Pli: Clothes that Grow

Petit Pli | 中国设计智造大奖佳作奖作品 | 2021
Petit Pli | DIA HONORABLE MENTION | 2021

航空工程师赖安·亚辛于2017年创立了Petit Pli（法语，意为小褶皱）。这是一家材料技术公司，创始人亚辛从他在伦敦帝国理工学院开发的可部署纳米卫星结构中汲取了灵感，通过数百个概念测试，开创了一种可生长服装，其工程设计已获得专利。

Petit Pli is a material technology company engineering patent accepted clothes that grow. Trained aeronautical engineer Ryan Yasin founded Petit Pli in 2017. Determined to pioneer a new approach, Ryan tested hundreds of concepts and drew inspiration from his background in deployable nano-satellite structures he developed at Imperial College London.

快精灵飞行器
Rapid Elves Aircraft

李培新 | 2023
Peixin Li | 2023

本产品整个机身以水滴作为设计灵感并展开外观设计，其造型可以在飞行器飞行时降低风阻，两个螺旋桨的外罩沿用了水滴形的设计，让整个飞行器的设计语言得以统一，机身骨架使用了重量轻、硬度高、耐磨性好的航空铝材制作而成。机身外壳、螺旋桨等则使用了昂贵的碳纤维复合材料，中间发动机支撑部分采用航空铝合金设计，总体结构做到了目前为止结构最牢固、重量最轻，合理的结构模式也增强了飞行器的经济性。其顶部由两个螺旋桨左右并列组成，直接控制飞行器的升降、转向，双螺旋桨自带的反方向旋转的工作原理直接平衡了反扭，填补了国内双螺旋桨无人机的制造空白。

The "Rapid Elves" aircraft draws design inspiration from the streamlined shape of a water droplet to minimize aerodynamic drag during flight. This design extends to the droplet-shaped propeller covers, ensuring a cohesive look. The frame is made of lightweight, high-strength aerospace aluminum, while the exterior and propellers are crafted from premium carbon fiber composites. The central engine support uses aerospace aluminum alloy, resulting in a structure that is both robust and lightweight, optimizing performance. The aircraft features two parallel propellers on the top for lift and directional control, which has filled the gap in domestic dual-propeller drone manufacturing. The counter-rotating propellers effectively balance torque, enhancing stability and efficiency.

曲水流觞：尺度的反思
Winding Scenery: The Reflection on Furniture Dimension

杨威杰 | 2020
Weijie Yang | 2020

一块黑色几何方碑出现在一群人猿前，这是库布里克《2001太空之旅》中的经典画面，这块黑色几何体象征着超越人类文明的力量，更象征着我们渴望获得的智慧。几何是我们解释世界的一种数学工具，更是我们所熟悉的一种生产方式，它代表着文明的尺度。简单的几何形态能够被快速地加工，以一种超自然的速度呈现在全世界的消费者面前，我们是不是该思考如此频繁出现的几何形态是否代表着文明的炫耀或者是消费文化的泛滥？曲水流觞椅所呈现的就是这样一种反思，让曲面作为形态的主体在空间中进行生长、发散并建构一种迥然不同的家具设计语言。

The black geometric monolith appearing before a group of apes in Stanley Kubrick's *2001: A Space Odyssey*, symbolizes a power that transcends human civilization and represents the wisdom we yearn for. Geometry serves as a mathematical tool for understanding the world and a familiar production method, embodying the scale of civilization. Simple geometric forms are quickly processed and presented to global consumers at an almost supernatural speed. This prompts us to question whether these frequently appearing geometric shapes signify a display of civilization or the proliferation of consumer culture. The "Winding Scenery" chair reflects this introspection, using curved surfaces as the main form to evolve, diffuse, and establish a distinctly different language in furniture design.

Orka 2：创新型助听器
Orka 2: Innovative Hearing Aid

齐思 | 2020
Si Qi | 2020

Orka 2是Hiorka与齐思联手设计的一款创新型助听器，旨在为听力受损的用户提供增强的听觉体验和生活质量。除了小巧的体积、精致舒适的设计，Orka 2通过先进的数字AI处理技术，根据用户个别的听力损失情况提供定制化的声音效果。主要功能包括消除啸叫的反馈消除、提高语音清晰度的降噪功能、可直接从智能手机和其他设备传输音频的蓝牙连接功能。Orka 2配备可充电锂电池，每次充电可使用 24 小时。

The Orka 2 is an innovative hearing aid designed by Hiorka and Si Qi to provide an enhanced listening experience and quality of life for hearing-impaired users. In addition to its compact size and comfortable design, Orka 2 uses advanced digital AI processing technology to provide customized sound effects according to the user's individual hearing loss. Key features include feedback cancellation to eliminate acoustic noise, noise cancellation to improve speech intelligibility, and Bluetooth connectivity to transmit audio directly from smartphones and other devices. Orka 2 is equipped with a rechargeable lithium battery that can be used for 24 hours per charge.

JetPods 无线蓝牙耳机
JetPods Wireless Bluetooth Earbuds

齐思 | 2023
Si Qi | 2023

齐思为国内蓝牙耳机品牌魔宴开发了JetPods蓝牙入耳式耳机。JetPods设计了一款专为年轻用户设计的"半入耳"耳塞,耳机的人体工程学设计细节提供更稳定舒适的长时间佩戴体验。通过可靠的音频技术和新的核心芯片降低了连接延迟,JetPods以实惠的价格实现了卓越的音质及无缝的操作佩戴体验。除了标志性的外观造型,设计中考虑的优化的分件数量和模块化元素有助于降低成本,并为未来型号提供了颜色多样性的可能。

JetPods features a "semi-in-ear" design tailored for younger users, with ergonomic details providing a stable and comfortable fit for extended wear. The headphones use reliable audio technology and a new core chip to minimize connection delays, offering excellent sound quality at an affordable price and seamless user experience. In addition to its distinctive appearance, the design includes optimized component numbers and modular elements, which help reduce costs and allow for future color variations.

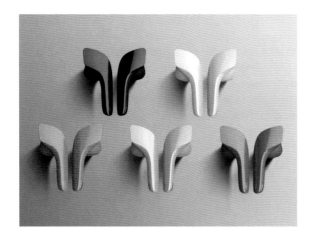

悠启超薄走步机
UREVO Treadmill

杨继栋 | 2022
Jidong Yang | 2022

悠启走步机,专为家庭与办公空间设计,适合利用碎片时间实现轻量有氧锻炼的各类人群,轻薄易收纳。它支持无线连接家居智能系统,轻松管理运动与健康数据。

UREVO Treadmill, designed specifically for home and office spaces, is ideal for those who want to engage in light exercise during fragmented time. It is lightweight, slim, and easy to store. Supporting wireless connection to smart home systems, it allows for effortless management of sports and health data.

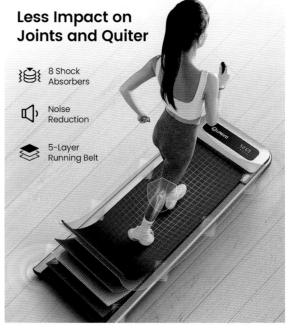

记忆之恒
Everlasting Memory

毛项杰 | 2022
Xiangjie Mao | 2022

设计师毛项杰是苏州河岸线长宁区、普陀区的设计团队主持人。行人漫步于苏州河海烟物流段，可见一个穿旗袍的女性牵着"永久"牌自行车的身影，同时这也是鲜活的雕塑座椅。"记忆之恒"这一作品印刻着永久自行车的标识变化过程，让行人得以寻找苏州河畔的历史印记，见证城市发展和时代变迁。作品在夜间的投影与周围环境互动的光影，动态体现了工业元素的律动。

The designer is the head of the design team for the Suzhou River bank line in Changning and Putuo districts. In the Suzhou River bank Haiyan section, people can see a sculptural lady in cheongsam holding a "Forever" bicycle, with sculptural seats aside. "Everlasting Memory" captures the evolving logo of the "Forever" bicycle, allowing the passers-by to search the historical traces along the Suzhou River, witnessing the urban development and changes over time. Nighttime projections and interactive light-shadow dynamics reflect the rhythm of industrial elements.

春语者：1号、2号、3号
Spring Messengers: 1, 2, 3

刘毅 | 2021—2023
Yi Liu | 2021-2023

"又来到了万物复苏的季节，各路使者们伴随着春风，带着时尚的生活和惬意的心情，带着新生的味道和绿色的观点，带着清新的花香和滋润的雨露，带着一份吉祥和美丽的祝福，来到了我们人间。他们就在我们身边……与我们一起工作，学习，娱乐和生活。给予我们春天最美好的信息和寄语。—— 致谢春语者"

"Once again, we arrive at the season of renewal. Various messengers, accompanied by the spring breeze, bring fashionable living and pleasant moods, the taste of new beginnings and a green perspective, the freshness of floral scents and nourishing rain, along with wishes of prosperity and beauty to our world. They are right here with us… working, learning, entertaining, and living alongside us, delivering the most beautiful messages and greetings of spring. — With thanks to the Spring Messengers."

鸟鸣电台
Birds Radio

刘毅 | 2022
Yi Liu | 2022

鸟鸣电台，起于2022年4月的上海新冠疫情期间，是一个由艺术家刘毅所发起的可持续公共艺术共创项目。他通过社交媒体向公众收集鸟鸣录音并且分享。至今已收集了来自上海、中国其他城市，以及包括世界各地的鸟鸣录音1000多个。意在"发现并分享身边本有的美好"。这也是疫情下所形成的共鸣与疗愈。目前，此项目还在持续进行与生长中……

Birds Radio began during the pandemic in Shanghai, April 2022. It is a sustainable public art co-creation project initiated by artist Yi Liu. Through social media, he has gathered and shared recordings of birdsong from the public. To date, over 1,000 bird songs have been collected from Shanghai, other cities in China, and around the world. The project's aim is to "discover and share the inherent beauty around us", reflecting a resonance and healing formed during the pandemic. The project is ongoing and continues to evolve.

墨石2
Inkstone 2

刘毅 | 2021
Yi Liu | 2021

本作品来源于日常的每日手机绘画，实现从虚拟网络的电子作品到物理空间的转换与再现。从雕塑和公共艺术角度而言，它是对中国传统的湖石"景观雕塑"的致敬，以及对当下城市语境的反思与突破。同时，作品也践行着艺术家对虚拟与现实的空间边界的探索，完成了网络绘画到现实物理空间的转换。自然与科技，古意与现代在此碰撞、融合。

This work originates from daily mobile phone drawings, transforming electronic works from the virtual network into physical space. From the perspectives of sculpture and public art, it pays tribute to traditional Chinese "landscape sculpture" of lake stones while reflecting and breaking through the contemporary urban context. It also embodies the artist's exploration of the boundaries between virtual and real space, achieving the transition from digital drawings to tangible physical space. This piece represents the collision and fusion of nature and technology, tradition and modernity.

展览设计
Exhibition Design

装配式策略被继续运用在展陈设计上。2023年，我们与"盒子社区"合作，用一个个盒子指代同济大学设计创意学院融入社区邻里的实验室群落。这些盒子与实验室出品的创新作品一起，共同模拟了这个嵌入城市的创意社区的活力和魅力。演绎了"上海这座城市就是最好的设计大学"的理念。

The prefabricated strategy continues to be applied in exhibition design. In 2023, we collaborated with the "Box Community" to use individual boxes to represent the laboratory community generated by Tongji University's College of Design and Innovation into the community neighborhood. These boxes, together with creative works produced by these laboratories, simulate the vitality and charm of this creative community embedded in the city. It embodies the concept that "Shanghai is the best design university".

主创构思：娄永琪

设计执行：郭泠

Creative Concept：Yongqi Lou

Design Execution：Ling Guo

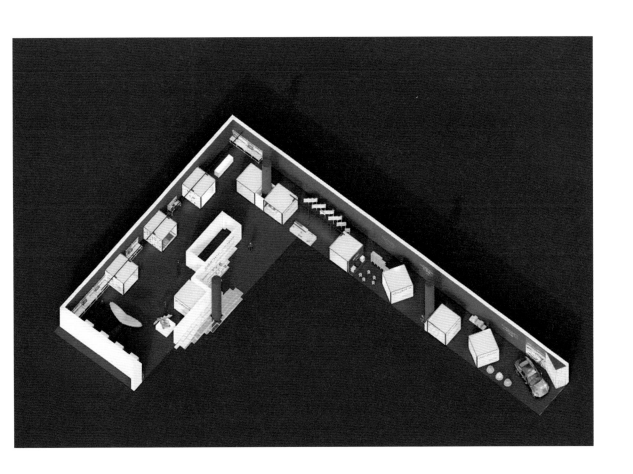

Exhibition Design

展览实景
Exhibition Scenes

Exhibition Scenes

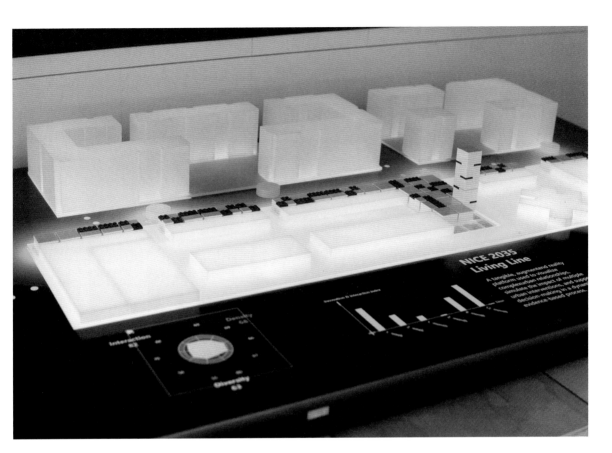

Exhibition Scenes

Exhibition Scenes

展览实景

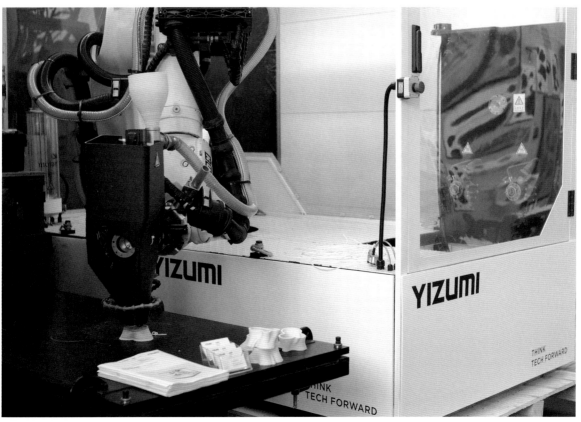

Exhibition Scenes

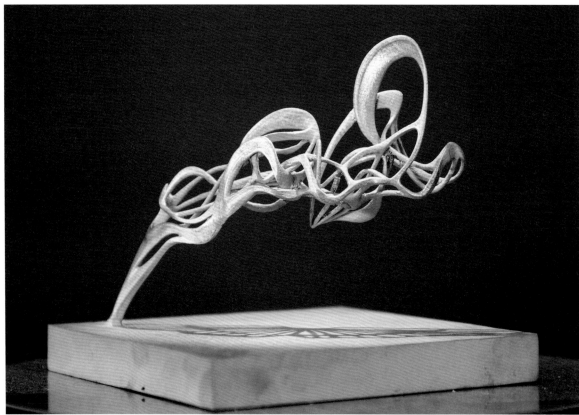

展览实景

工作坊
Workshops

一块砖 / Just One Brick

学术支持: 祎设计

特别支持: 安东·西比克

活动简介: 祎设计将工业陶瓷废料转化为可用于装饰、结构、且可定制的各类瓷砖、建筑砖,为室内和建筑行业提供可循环选项。我们的实验室研发了一种由回收陶瓷废料为主要原料的新型材料。在首次使用这种新材料制作砖和瓷砖之后,许多其他创意用途也随之而来。仅用一块砖,我们就可以制作胸针、杯子、桌子、雕塑、喷泉和灯具等。"Just One Brick"是让广大建筑师、产品设计师、艺术家都能参与进来尝试制作的项目。目的是强调使用这种陶瓷再生材料的创新可能性,为社区和我们的星球共同推进设计。同时在这次展会中,我们也展出了不同的产品,包括由陶瓷废料和其他工业固废制成的100%回收砖。

Partners: Yi Design

Support: Aldo Cibic

About: Yi Design offers a circular recycled product for the interior and architecture industry, transforming industrial ceramic waste into decorative, structural, and customizable tiles and bricks, art and product design. Yi Design has invented and developed a material made from recycled ceramic waste. After first making bricks and tiles with the new material, many other creative uses have emerged. From just one brick, we have made brooches, cups, tables, sculptures, fountains, and lighting. Architects, product designers and artists are invited to participate in this "Just One Brick" project. The project aim is to highlight the innovative possibilities using this recycled brick material for themselves, advancing design for the community, and for our planet. In this exhibition, the different products are showcased, including 100% recycled brick made out of porcelain waste and other wastes.

做你自己的……Whatchamacallit?

学术支持: Fablab O Shanghai数制工坊

特别支持: 萨维里奥·西利

活动简介: 此次工作坊是一次动手之旅,体验如何将感官观察与制作制造有效结合起来,在超越人类的语境里开展设计研究。

在"以人为本"的设计实践中,观察和访谈是重要的研究部分,帮助我们与设计对象建立共情,探索意想不到的洞察。当设计领域和设计挑战转变为"以自然为本"时,我们又该如何建立共情、探索洞察呢?

此次工作坊是一次实验,探究如何运用设计和制作制造来丰富我们探索自然的感官。我们将使用经济易得的材料和实物构思的技巧来制作探索自然的工具:它们的功能可能并不显而易见,也可能没有专门的名称,所以……我们……叫它们"Whatchamacallit"。

Make Your Own... Whatchamacallit?

Partners: Fablab O Shanghai

Support: Saverio Silli

About: This workshop is a hands-on journey to experience how sensorial observation and fabrication can be effectively combined to conduct design research in a more-than-human context.

In the Human Centred Design practice, observations and interviews are a crucial part of the research phase to establish empathy with the people we are designing for and explore new and unexpected points of view. How can we produce the same results when dealing with Nature Centred domain and challenges?

The workshop is an experiment of how we can use design and fabrication to enrich our senses in the exploration of Nature. We will use cheap materials and physical ideation techniques to fabricate our personal exploration tools: their function might not be obvious and they might not have a dedicated name, so... "whatchamacallit"?

造物/材料工作坊　　OrganiX Workshop

学术支持: 造物实验室, 材料与应用创新实验室

特别支持: 伊之密

活动简介: 造物/材料工作坊以讲座及材料和加工认知为主,首先介绍有机材料大类及典型材料特征,通过材料小样给予感官体验,其次通过三维设备展示增材制造方式,最后介绍体验展位展出的部分展品。

Partners: Making Lab, Material and Application Design Laboratory

Support: YIZUMI

About: OrganiX workshop focus on introduce the organic category and the typical and special orgainc material type and their character and performance. By the detail introduce of large sized 3D printer and its working principle, present the exhibits and their innovation aspects.

AIGC工作坊

学术支持：设计人工智能实验室

特别支持：特赞科技

活动简介：AIGC工作坊鼓励参与者手脑并用，发挥创意。本次工作坊旨在通过基础知识讲解和实践，为学员带来生动的AI绘画体验。

内容包括：AI绘画的基本概念和方法、提示结构和优化逻辑、如何使用ControlNet和Lora等。AIGC工作坊使用的工具"MuseAI"由特赞科技提供。如需更多体验，请访问 musai.cc。

AIGC Workshop

Partners: Design A.I. Lab

Support: Tezign

About: The "AIGC Workshop" encourages the participants to use both their hands and brains to be creative. Through explanations of basics and practice, this workshop aims to bring a vivid AI drawing experience to the participants. The contents include basic concepts and methods of AI painting, prompt structure and optimization logic, how to use ControlNet and Lora, etc. The tool "MuseAI" used in the AIGC workshop is provided by Tezign Technology. For more experience, please visit museai.cc.

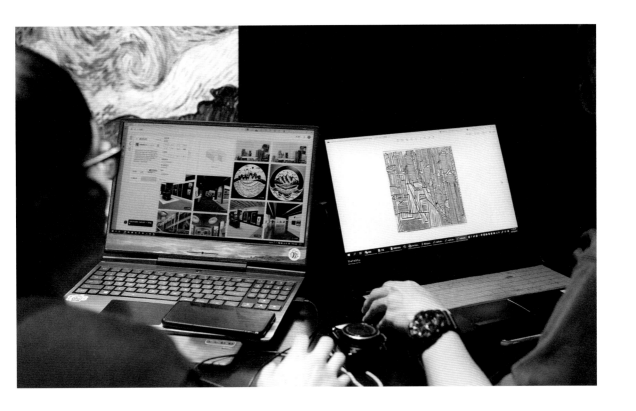

Bake工作坊   Bake Workshop

学术支持：机械臂实验室

特别支持：库卡（中国）机器人

活动简介：本次工作坊以对话方式开始，运用AI生成数字艺术图像，并使用机械手臂将数字作品转换为物理实体。参与者以小组为单位，通过多种媒介（工具头）呈现实体作品不一样的质感。鉴于大脑与肢体相互协调过程，AI与机器手臂进行互动，以直观的方式展现新的艺术形式和价值创新。机械臂作为数字与物理世界的桥梁，使创作过程变得更具多样性与真实性。

Partners: Robotic Arm Lab

Support: KUKA

About: This workshop started with a dialogue, using AI to generate digital art images, and converting digital works into physical entities by robots. Participants work in groups and use a variety of media tools to present different textures of their works. As coordination process between the brain and the limbs, AI interacts with the robots to display new art forms and value. As a bridge between the digital and physical worlds, robotic arms make the creative more diverse and realistic.

SoCity Community DAO工作坊

学术支持：同济-麻省理工上海城市科学实验室

特别支持：SoCity，社交图层

活动简介：工作坊探讨如何通过SoCity Community DAO这一去中心化组织，建立一个能够借助去中心化激励政策，促进城市亲社会、可持续行为发生的生态系统。在工作坊中，参与者了解了围绕区块链、智能合约、代币经济、去中心化治理、城市模拟和算法分区的技术和研究工作。然后以小组的形式，使用SoCity团队开发的网页端城市交互模拟工具，探索新的、亲社会的、去中心的城市发展过程。

SoCity Community DAO Workshop

Partners: Tongji-MIT City Science Lab @Shanghai

Support: SoCity, Social Layer

About: The discussion was based on SoCity, a non-profit decentralized autonomous organization (DAO) that aims to establish an ecosystem that promotes prosocial, pro-sustainability behaviors in cities with decentralized incentive policies. During the workshop, the participants were introduced to the technologies and research works around blockchain, smart-contract, token economies, decentralized governance, urban simulations, and algorithmic zoning. In groups, they used the website-based urban simulation tool developed by SoCity team, to explore novel, prosocial, decentralized urban development processes.

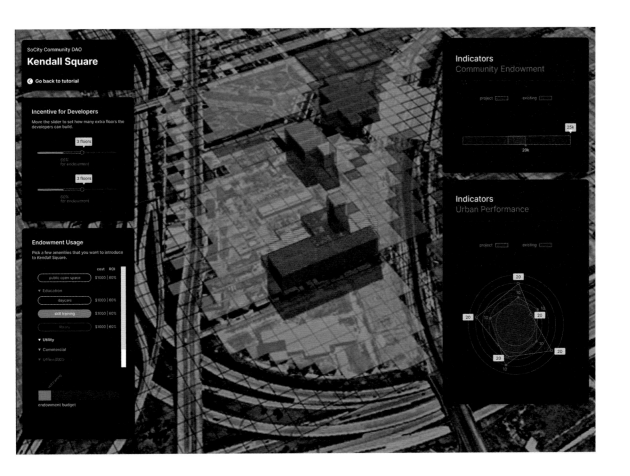

食物工作坊 / Food Workshop

学术支持：好公社

特别支持：哈巴庄园、野酵

活动简介：

Part 1: 正念饮食 x 哈巴农场

尽管正念饮食正变得很流行，但是大多数人并不知其所以然。当我们吃一个苹果时，我们会注意苹果对我们的五感分别带来什么感受吗？还是我们只需要慢慢吃东西？正念饮食就是慢慢吃，这样简单吗？

Part2: 大米味噌—发酵的智慧 x 野酵

味噌是一种比酱油更古老的发酵食物，在这个高速高科技的时代，等待食物发酵成美味已然是一份奢侈，静下心来，给自己一点时间参与到食物的生命周期中，学会珍惜和利用好每一份食物。

Partners: NICE COMMUNE

Support: Haba Manor, Wild Fermentation

About:

Part1: Mindful Eating x Haba Farm

While mindful eating is becoming increasingly popular, most people do not understand why. When we eat an apple, do we pay attention to what sensations the apple brings to our senses? Or do we simply eat slowly? Is mindful eating really that simple?

Part 2: Rice Miso - The Wisdom of Fermentation x Wild Fermentation

Miso is a fermented food that is older than soy sauce. In this fast-paced and high-tech era, waiting for food to ferment into deliciousness has become a luxury. Take a moment to calm down and give yourself some time to participate in the life cycle of food. Learn to cherish and make good use of every meal.

HDPE循环再造首饰工作坊 | HDPE Recycling Jewelry Workshop

学术支持：JALAB首饰实验室

特别支持：阿普塔

活动简介：塑料包装盒等产品残留物的归宿是什么？

我们能否通过什么方式赋予残破的塑料二次生命？

让工作坊老师带你一起体验塑料的循环再造。通过创意将HDPE板材制作成为一个独一无二的首饰作品。

运用各种工具，充分发挥表达能力与动手操作能力，将压制回收塑料板制作成各式各样的首饰作品。

Partners: JALAB

Support: Aptar

About: What is the destination of product residues such as plastic packaging boxes?

In what ways can we give broken plastic a second life?

Let the workshop teacher take you through the recycling of plastic. Through creativity, the HDPE plate is re-made into a unique jewelry work.

Using a variety of tools, give full play to the ability of expression and hands-on operation, the pressed recycled plastic version is made into a variety of jewelry works.

致谢
Acknowledgements

同济大学	Tongji University
同济大学设计创意学院	College of Design and Innovation, Tongji University
杨浦区&四平路街道	Yangpu District & Siping Road Subdistrict
好公社	NICE COMMUNE
盒子社区	Box Community
Fablab O Shanghai数制工坊	Fablab O Shanghai
安东·西比克工作室	Aldo Cibic Studio
材料与创新应用实验室	MA&D
SustainX 可持续未来设计研究中心	SustainX Lab
尚想实验室	Shang Xiang Lab
社交图层	Social Layer
声音实验室	Sound Lab
时尚中心	FINE Center
JALAB首饰实验室	JALAB
同济-阿斯顿·马丁创意实验室	Tongji-Aston Martin Creative Lab
同济-麻省理工上海城市科学实验室	Tongji-MIT City Science Lab @Shanghai
机械臂实验室	Robotic Arm Lab
设计人工智能实验室	Design A.I. Lab
辛思想设计思维与战略咨询	XXY Innovation
祎设计	Yi DESIGN
载运工具与系统创新设计实验室	Next Mobility Lab
造物实验室	Making Lab
张周捷数字实验室	Zhang Zhoujie Digital Lab
中国设计智造大奖	DIA Award

WDCC 设计花絮
Design Sidelights of WDCC

世界设计之都大会
主视觉设计

Main Visual Identity System Design of WDCC

2022—2023年世界设计之都大会主视觉设计由同济大学设计创意学院王敏教授和杜钦副教授领衔设计，由资深设计师李梦媛和来自纽约的Karlssonwilker设计事务所进行了延展设计。王敏教授认为："主视觉设计应当增强城市的肌理与层次，让城市更具律动和活力，连接城市空间，促进人与人沟通。同时，主视觉设计的形态应该从都市的形态而来，展现都市的美好与活力。"大会标志由英文首字母WDCC、彩色图形元素，以及大会中英文全称组成。在设计应用中，我们将大会主视觉与上海城市天际线、建筑空间，以及各种设计发生的场景进行了并置，形成视觉语言上的动态关联，呈现设计的动能，创意的活力，寓意设计高度融入城市生活。

The Main Visual Identity System Design of WDCC 2022-2023 was led by Prof. Min Wang and Assoc. Prof. Qin Du from the College of Design and Innovation, Tongji University. The design was further developed by senior designer Mengyuan Li and the New York-based design firm Karlssonwilker. Prof. Wang believes that "the main visual design should enhance the texture and layers of the city, bringing more rhythm and vitality to it, connecting urban spaces, and facilitating communication between people. The form of the main visual design should be derived from the urban form, showcasing the beauty and vibrancy of the city." The logo is composed of the acronym WDCC, colored graphic elements, and the full name in both Chinese and English. In its application, the main visual design is juxtaposed with Shanghai's skyline, architectural spaces, and various design-related scenes to create a dynamic visual language that reflects the energy of design and the vibrancy of creativity, symbolizing the deep integration of design into urban life.

A1
彩色图标中的线条
作为辅助图形

A2
彩色图标+
英文字标组合

A3
彩色图标+
中英文字标组合

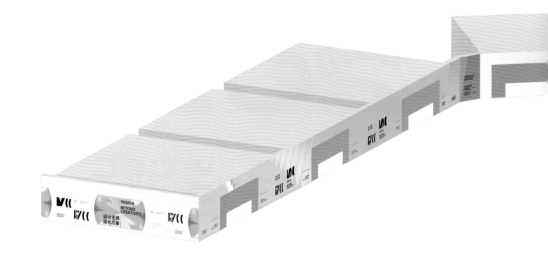

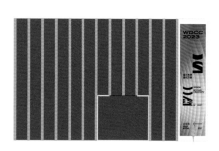
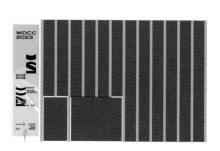

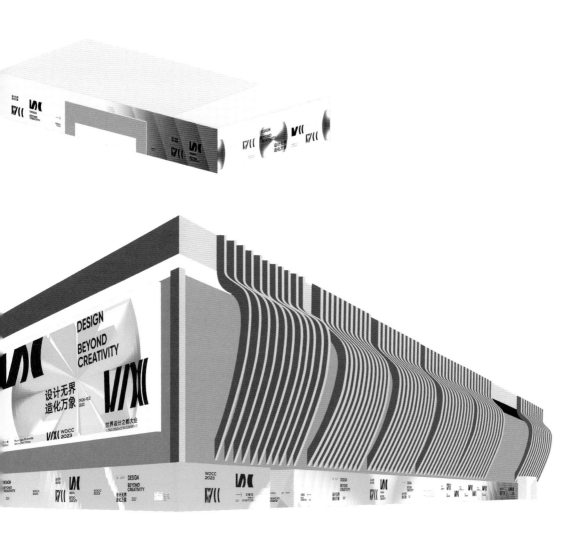
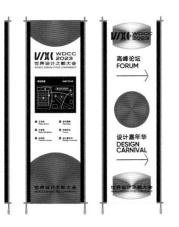

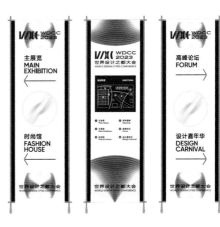

Design Sidelights of WDCC • Main Visual Identity System Design of WDCC

世界设计之都大会开幕秀：设计交响（2022）

Opening Show of WDCC 2022: Design Symphony

2022年世界设计之都大会开幕秀数字设计交互作品《设计交响》由同济大学设计创意学院媒体与传达团队和声音实验室主创。本作品通过沉浸式影像、灯光、舞蹈表演展现如"交响诗"般的视听体验，运用科技手段，融入算法生成设计、动作捕捉与3D扫描技术，颂扬设计对产业、城市和人所作出的可见贡献。三位专业舞者身着惯性动作捕捉设备，通过丰富而优美的身体动作来驱动数字算法3D生成动态影像，诗意画面配合现场270度环幕的全景视觉观赏体验以及灯光设计，分别展现了设计对产业创新、城市活力提升与服务系统优化的驱动力，呼应大会"设计无界，相融共生"的主旋律。

The WDCC 2022 opened with a digital design interactive performance titled *Design Symphony*, created by the Media and Communication division and the Sound Lab of the College of Design and Innovation at Tongji University. This piece offered a symphonic audiovisual experience through immersive imagery, lighting, and dance performances. Utilizing technology, it incorporated algorithmic design, motion capture, and 3D scanning techniques to celebrate the visible contributions of design to industry, cities, and people. Three professional dancers, equipped with inertial motion capture devices, used their rich and graceful movements to drive digital algorithms that generated dynamic 3D visuals. The poetic imagery, combined with a 270-degree panoramic viewing experience and lighting design, showcased the driving force of design in industrial innovation, urban vitality enhancement, and service system optimization. This resonated with the conference's main theme, "Design Beyond Border, Diversity and Togetherness".

总策划：娄永琪

导演：张屹南、柳喆俊

舞台监督：李悦娟

设计开发：张屹南、柳喆俊、吴昱、石峰、杨书锋、郁新安、冯泗衡、谭丞超、刘文艺、刘隽语、黄秋韵、田梓昂

编舞：高腾

舞蹈表演：高腾、单阔阔、孟琪

音乐创作：约翰·尼格尔

灯光设计：陈民安

摄像：武洲

Chief Planner: Yongqi Lou

Director: Yinan Zhang, Zhejun Liu

Stage Manager: Yuejuan Li

Designed by: Yinan Zhang, Zhejun Liu, Yu Wu, Feng Shi, Shufeng Yang, Xin'an Yu, Siheng Feng, Chengchao Tan, Wenyi Liu, Junyu Liu, Qiuyun Huang, Zi'ang Tian

Choreographer: Teng Gao

Choreography: Teng Gao, Kuokuo Shan, Qi Meng

Music Composition: Johann Niegl

Lighting Design: Benjamin Chen

Videographer: Zhou Wu

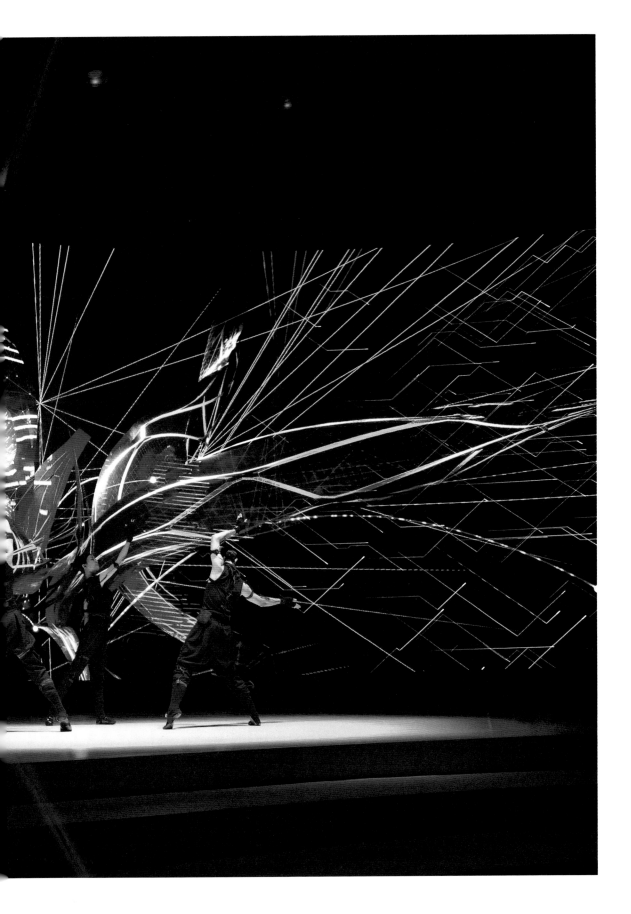

世界设计之都大会开幕秀：化（2023）

Opening Show of WDCC 2023: Metamorphosis

2023年世界设计之都大会开幕式新媒体演出作品《化》，由同济大学设计创意学院声音实验室主创，联合上海戏剧学院舞蹈学院共同呈现。演出内容从日常物件到社会图景、从科技前沿到自然生态，以充满诗意的表达演绎了设计在城市文化塑造过程中贡献的力量，并展望了智能时代设计的进化蜕变。本作品通过AI与算法技术将这些混搭式的创新产出整合为舞台上的综合体验，并努力尝试向自然、生态与人文回归，呈现了智能时代设计文化的一种未来可能性，以及运用中国智慧让设计焕发新生的美好愿景。

The new media performance piece "Metamorphosis" was presented at the opening ceremony of WDCC 2023. It was created by the Sound Lab of the College of Design and Innovation at Tongji University, in collaboration with the Shanghai Theatre Academy. The performance explored themes ranging from everyday objects to societal landscapes, from cutting-edge technology to natural ecology, using poetic expression to showcase the power of design in shaping urban culture and to envision the evolution and transformation of design in the intelligent era. This piece integrated these diverse, innovative outputs into a comprehensive stage experience using AI and algorithmic technology, while striving to reconnect with nature, ecology, and humanity. It presented a possible future for design culture in the era of AI, embodying the vision of revitalizing design with Chinese wisdom.

总策划：娄永琪

导演：张屹南

执行导演：李悦娟

舞蹈编创与表演：马惠霞、牛梦涵、张乐乐、李志杰、杨英琪

原创音乐：约翰·尼格尔

设计开发：张屹南、吴昱、石峰、温子阳、谭丞超、冯泗衡、刘文艺、张馨月、刘隽语、刘一驰、付雨彤、施绿筠、岳洪宇、王烽宇、张欧琦、李悦娟、金怡、魏有晟、杨书锋、顾天润、胡琦轩

灯光设计：陈民安

舞台监督：刘晶、邓晴

古筝演奏：刘羽云

伴唱：邓晴

摄影摄像：武洲、林慈丰、林尤翔

纪录片制作：张屹南

项目助理：刘曌

特别感谢：上海戏剧学院 张麟

Chief Planner: Yongqi Lou

Director: Yinan Zhang

Executive Director: Yuejuan Li

Choreography and Performance: Huixia Ma, Menghan Niu, Lele Zhang, Zhijie Li, Yingqi Yang

Original Music: Johann Niegl

Designed by: Yinan Zhang, Yu Wu, Feng Shi, Ziyang Wen, Chengchao Tan, Siheng Feng, Wenyi Liu, Xinyue Zhang, Junyu Liu, Yichi Liu, Yutong Fu, Lüyun Shi, Hongyu Yue, Fengyu Wang, Ouqi Zhang, Yuejuan Li, Yi Jin, Yousheng Wei, Shufeng Yang, Tianrun Gu, Qixuan Hu

Lighting Designer: Benjamin Chen

Stage Manager: Jing Liu, Qing Deng

Guzheng: Yuyun Liu

Accompanied by: Qing Deng

Photography: Zhou Wu, Cifeng Lin, Youxiang Lin

Documentary Production: Yinan Zhang

Project Assistant: Zhao Liu

Special Thanks: Lin Zhang, Shanghai Theater Academy

世界设计之都大会"人民城市 | 处处有设计"建筑展（2023）

"People's City | Design Everywhere" Architectural Exhibition of WDCC 2023

"人民城市 | 处处有设计"建筑展位于 2023 年世界设计之都大会主展区，融合体验、学习和探索，聚焦于城市更新与全球智慧、智能设计与数字建造、建成环境与遗产保护、建筑城市空间一体化、未来城市建筑实验室等议题，从全系谱视角展现多维度城市更新背景下的城市建筑创新实践、历史传承和全球智慧。通过国内外设计赋能高质量发展、高品质生活、高效能治理的案例，诠释"城市处处有设计，生活处处有设计"的创新融合发展观。

The "People's City | Design Everywhere" architectural exhibition is located in the main exhibition area of the WDCC 2023. It integrates experience, learning, and exploration, focusing on topics such as urban renewal and global intelligence, smart design and digital construction, the built environment and heritage conservation, the integration of architecture and urban space, and future urban architectural laboratories. The exhibition showcases innovative practices in urban architecture, historical preservation, and global knowledge in the context of multidimensional urban renewal from a comprehensive perspective. Through case studies of domestic and international design empowering sustainable development, high-quality living, and efficient governance, it interprets the innovative and integrated development concept that "design is everywhere in the city, and everywhere in life."

策展人：李翔宁、汤朔宁、刘刊

主办单位：同济大学

联合主办单位：中国建筑学会建筑评论学术委员会

Curators: Xiangning Li, Shuoning Tang, Kan Liu

Host: Tongli University

Co-host: Architectural Review Academic Committee of the Architectural Society of China

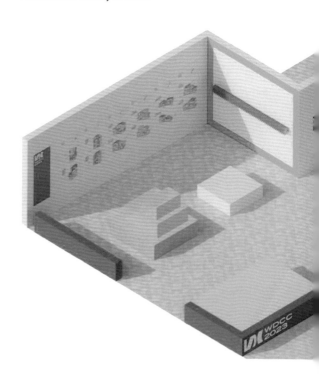

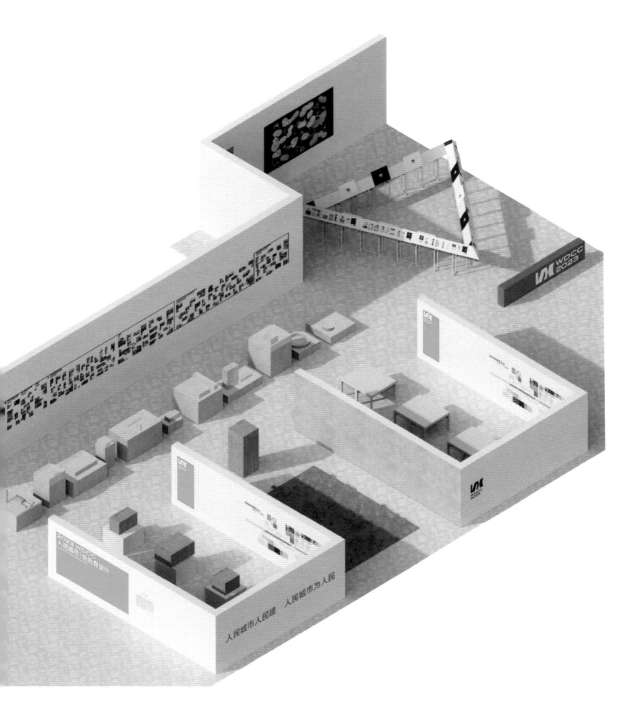

世界设计之都大会会场设计（2022—2023）

Site Planning of WDCC 2022-2023

2022—2023 年世界设计之都大会场地总体规划设计由同济大学章明教授领衔，其对于大型公共空间设计和公共活动配套建构筑物的设计具有丰富的经验，曾担任杨浦滨江南段公共空间、苏州河黄浦区段公共空间总设计师及第三届上海城市空间艺术季总建筑师。团队成员还包括同济设计集团·原作设计工作室设计总监张姿，副主任建筑师丁纯，建筑师郭璐炜、吴炎阳。

2022 年大会的会场设计通过三大公共空间艺术装置——"设计叙事""未来之眼"和"设计之梦"来构建独特的空间体验。会场运用了前沿设计思维与创新技术，将大会公共空间点亮。同时，会场设计也与主展陈设计相融合，为主旨内容的呈现提供了展示载体。"设计叙事"位于主展入口，以火炬造型，齐聚 43 个世界设计之都城市；"未来之眼"位于主展中心，打造充满科技感的元宇宙空间；"设计之梦"位于主展馆和主会场中间的广场，以纯真梦幻的造型，放飞设计梦想。

2023 年大会主入口形象装置设计以"无界之塔"为概念，取义"无界之境中的灯塔"。设计贯彻可持续再生的原则，实现经济、高效的搭建模式，利用场地既有元素，采用轻介入的建构方式，以脚手架构成的"三棱柱"横亘于二层桥体之上，10m 高的塔身以虚透的边界融于天空，营造出开放融通、生长于场所中的无界之塔。团队在整体场地规划中合理组织场地流线，把控会场氛围营造，以彩绘地贴的形式将大会主视觉地景化，同时兼顾片区划分与路径引导。

The overall planning and design of the 2022-2023 World Design Capital Conference venue was led by Prof. Ming Zhang from Tongji University. He has extensive experience in designing large public spaces and supporting structures for public activities. He previously served as the chief designer for the Yangpu Riverside South Section public space, the Huangpu District public space of the Suzhou Creek, and the chief architect for the 3rd Shanghai Urban Space Art Season. The team also included Zi Zhang, Design Director of the Original Design Studio at Tongji Design Group; Chun Ding, Deputy Chief Architect; and architects Luwei Guo and Yanyang Wu.

The site planning for WDCC 2022 offers a distinctive spatial experience through three major public art installations —"Design Narrative", "Eye of the Future" and "Design Dream". By incorporating cutting-edge design thinking and innovative technologies, the site planning illuminates the public spaces of the event. It seamlessly integrates with the main exhibition, providing a dynamic platform for presenting the conference's key themes. "Design Narrative", located at the exhibition entrance, features a torch-like structure symbolizing the gathering of 43 World Design Capital cities. "Eye of the Future", at the heart of the exhibition, creates a tech-driven metaverse-inspired space. "Design Dream", positioned in the plaza between the exhibition hall and the main venue, with its whimsical design, symbolizes the limitless aspirations of design.

The design of the main entrance installation for the 2023 conference was conceptualized as the "Tower of Boundlessness", symbolizing a "lighthouse in a boundless realm." The design adheres to principles of sustainable regeneration, achieving an economical and efficient construction mode. It utilizes existing elements of the site and employs a light intervention approach. A "triangular prism" structure made of scaffolding is placed horizontally above the second-level bridge. The 10-meter-high tower integrates into the sky with its transparent boundaries, creating an open and connected "Tower of Boundlessness" that grows within the site. The team organized the site's circulation paths reasonably, controlled the atmosphere of the venue, and used painted ground stickers to integrate the main visual elements of the conference into the landscape. The area division and path guidance are also taken into account.

世界设计之都大会会场设计（2022—2023） WDCC设计花絮

2022年WDCC户外广场因地制宜布置的艺术作品
Outdoor artworks in the plaza WDCC 2022, arranged according to local conditions

世界设计之都大会主题馆：无止园（2022）
WDCC 2022 Theme Pavilion: Endless Garden

无止园是2022年世界设计之都大会主题展的重要组成部分，是所有主题馆工作坊的举办场所。脚手架主体框架和聚碳酸酯板采光顶由标准固定件和拉索连接，其空间形态设计转译于东方意韵：从传统屋脊的连绵错叠到边界的起伏消隐，从片墙窄巷的穿行到弥漫路径的游走，从落雨听风的黛瓦幽园到阳光穿隙的鳞鳞空境，"园"作为传统空间形式得到了全新演绎，在移步易景中赋予了"无止"的感受。与此同时，设计师提供了镶嵌在脚手架框架之中的可以登高环游建筑的体验路径，体现了公共空间的开放度。

场景照片由章明教授提供

The "Endless Garden" was an essential part of the theme exhibition for WDCC 2022 and served as the venue for all the theme pavilion workshops. It featured a main framework made of scaffolding and a polycarbonate light-transmitting roof connected with standard fixtures and still cables. The spatial design translates elements of Eastern aesthetics: from the overlapping ridges of traditional roofs to the undulating and disappearing boundaries, from the narrow passages of walls to the wandering paths, from the serene garden of dark tiles listening to the rain and wind to the sunlit gaps creating shimmering spaces. The "garden" as a traditional spatial form was reinterpreted, providing an experience of "endlessness" as one moves through different perspectives. Meanwhile, the designer offered an experiential path integrated into the scaffolding framework, allowing visitors to ascend and traverse the structure, reflecting the openness of the public space.

Photos provided by Prof. Ming Zhang

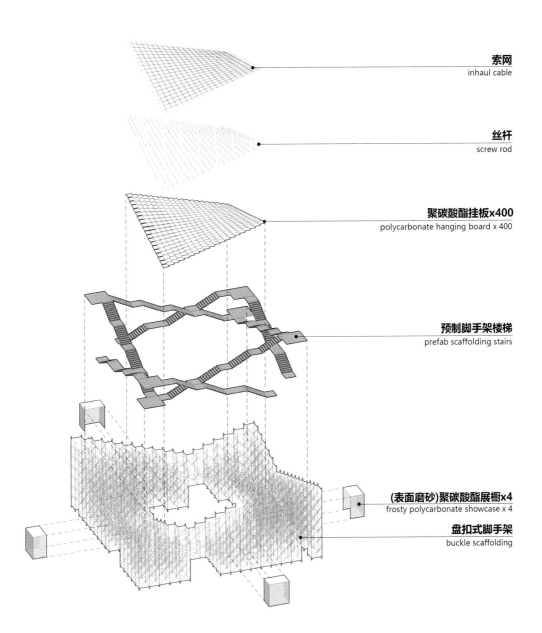

世界设计之都大会论坛场景设计（2023）

Scenography Design of the Conference Venue of WDCC 2023

设计是服务于真实世界的。

真实世界是一个复杂的有机系统。在这里，万物共享阳光雨露，互相影响，又各自繁衍，生生不息。

真实世界，是我们共同的家园。

设计无界，关怀所有的生命。

造化万象，可持续的发展才能创造共同的永恒的福祉。

据此，我们以"家园"作为 WDCC 2023 会场的场景创意概念。

2023 年世界设计之都大会场景设计的最重要的考量是绿色设计。这里有两层含义，首先是尽量采用装配式，最大限度地减少垃圾的产生；其次，希望把苗圃和鱼缸搬进会场，呈现人与动植物的共存。由脚手架、叉车板、散布的大小屏幕等构成会场的形式元素，配以轻松的陈设设计和灯光氛围，创造了全新的会议体验。

Design is for a real world.

Real world is a complex organic system, where everything shares the sun and rain, influences each other, flourishes and multiplies.

This real world is our home.

Design Beyond Borders, is caring for all lives.

Design Beyond Creativity, is to create our eternal well-being through sustainable development.

Thereby, we prompt the idea "Our Homeland" as the creative concept for this WDCC 2023 venue.

The key consideration for the scenography design of the 2023 World Design Cities Conference is "Green", which has two meanings. Firstly, try to use assembly solutions as much as possible to minimize the generation of garbage; secondly, we hope to bring plant nurseries and fish tanks into the venue to showcase the coexistence of humans, animals, and plants. The elements of venue are composed of scaffolding, forklift boards, scattered screens of various sizes, etc., combined with relaxed display design and lighting atmosphere, creating a new conference experience.

主创构思：娄永琪

执行负责：新绿墨、陈民安

Creative Concept: Yongqi Lou

Design Execution: New Green Ink, Benjamin Chen

设计是服务于真实世界的。

真实世界是一个复杂的有机系统。
在这里，万物共享阳光雨露，
互相影响，又各自繁衍，生生不息。

真实世界，是我们共同的家园。

设计无界，关怀所有的生命；
造化万象，可持续的发展才能创造共同的永恒的福祉。

据此，我们以"家园"作为
WDCC2023会场的场景创意概念。

创意概念 ： 娄永琪
设计执行 ： 新绿墨

世界设计之都大会演讲台设计

Design of the WDCC Main Podium

主演讲台是整个大会重要的品牌标识之一,也是一个不太容易设计的任务。在否定了一堆设计之后,策展人决定亲自出手设计。最终的方案是由大会 WDCC 标识演化拉伸而成,有点建筑的意味。整个演讲台由 3D 打印而成,方便音频线缆可以从中穿过。由于整体非常坚固、轻盈,故可以轻松地在弧形轨道上滑行。

The WDCC main podium is one of the most important parts of the visual identity system of the entire conference. At the same time, it's a task that is not easy to achieve. After rejecting a series of proposals, the curator decided to take on the challenge myself as the Creative Director of WDCC. The final design was evolved and extended from the conference's logo, with a hint of architectural elements. The entire podium was 3D printed, allowing audio cables to pass through it. Due to its solid yet lightweight structure, it can easily glide automatically on a curved track.

主创构思:娄永琪

设计执行:郭泠

3D 打印工程制作:周洪涛

Creative Concept: Yongqi Lou

Design Execution: Ling Guo

3D Printing and Engineering: Hongtao Zhou

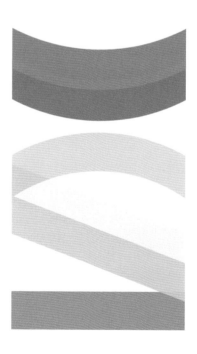

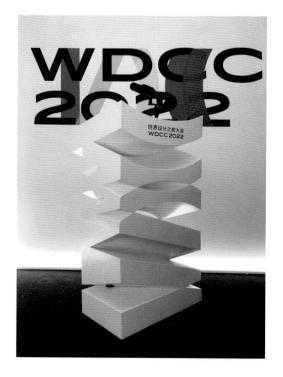

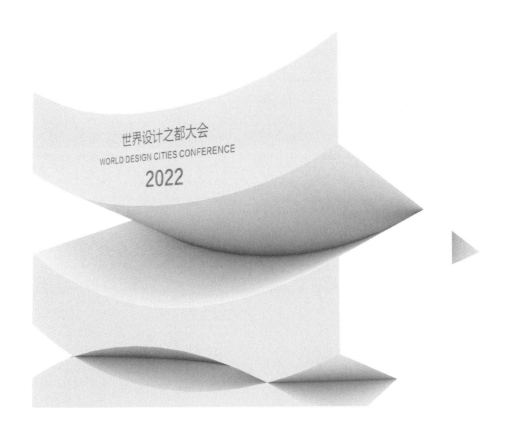

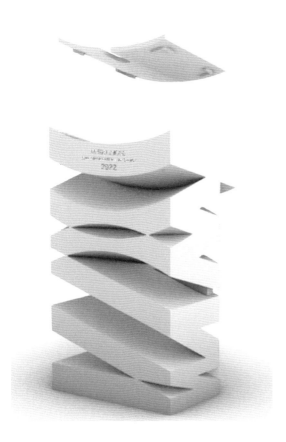

Design Sidelights of WDCC Design of the WDCC Main Podium

图书在版编目（CIP）数据

设计无界：世界设计之都大会主题展 / 娄永琪编著．
上海：同济大学出版社，2024.9．-- ISBN 978-7-5765-1348-6

Ⅰ．TU984-64

中国国家版本馆 CIP 数据核字第 2024JQ2719 号

本书以展览图册的形式，呈现了 2022 年和 2023 年在上海召开的世界设计之都大会的两届主题展的入选作品，及策展人对未来设计发展和上海设计之都建设的思考。2022 年及 2023 年主题分别为"设计无界，相融共生"和"设计无界，造化万象"。作者为这首两届大会牵头策展了两个相互关联的主题展，分别以"人·人，人·自然"和"创意社群：嵌入式、可感知与高交互"为题演绎大会主题。本书作为一本有价值的资料性图书，可为设计和创意人士提供参考。

This book, presented as an exhibition catalog, features selected works from the theme exhibitions of the World Design Cities Conference held in Shanghai in 2022 and 2023, along with curators' reflections on the future of design development and the construction of Shanghai as a Design Capital. The themes for 2022 and 2023 were "Vision in Perspective: Design Beyond Borders, Diversity and Togetherness" and "Design Beyond Creativity" respectively. The editor led the curation of two interrelated theme exhibitions for these conferences, titled "People to People, Human to Nature" and "NICE Commune: Embedded, Sensible and Interactive", interpreting the overarching themes of the conferences. As a valuable reference book, it aims to provide insights for design and creative professionals.

设计无界：世界设计之都大会主题展

娄永琪 编著

责任编辑　徐　希　周原田
责任校对　徐逢乔
装帧设计　张心怡
排版支持　王依琳　曹沛晴　何鸿禧　李谦谦
　　　　　张　曼　刘叶楠　李子祺　谢怡华

出版发行　同济大学出版社
　　　　　（地址 上海市四平路 1239 号　邮编 200092
　　　　　电话 021-65985622）
经　销　全国各地新华书店
印　刷　上海雅昌艺术印刷有限公司
开　本　710mm×1000mm 1/16
印　张　21
字　数　419000
版　次　2024 年 9 月第 1 版
印　次　2024 年 9 月第 1 次印刷
书　号　ISBN 978-7-5765-1348-6
定　价　288.00 元

本书为上海文化发展基金会图书出版专项基金资助项目

本书中项目作品图片均来自项目的提供方，会场实景照片来自 WDCC 官方摄影集，均已获版权方授权

本书封面来源于 WDCC 2023 大会主视觉，原设计由纽约 Karlssonwilker 设计工作室与同济大学设计创意学院团队联合完成

版权所有 侵权必究
印装问题 负责调换